# I Married
# A Dinosaur

# I Married A Dinosaur

## Lilian Brown

**COACHWHIP PUBLICATIONS**

Landisville, Pennsylvania

*I Married a Dinosaur*, by Lilian Brown
Copyright © 2010 Coachwhip Publications
First publication 1950.
No claims made on public domain material.


ISBN 1-61646-028-8
ISBN-13 978-1-61646-028-0

Back Cover: AMNH Hall of Dinosaurs (CC) Thomas Cowart

CoachwhipBooks.com

# CONTENTS

I never wrote an introduction to a book with more pleasure. "Pixie" was the name Barnum Brown used for his wife, and after you have read her story you will agree it was well chosen. Here is real humor, very good writing and the revelation of what it means to be a fossil hunter's wife. I've had a laugh from every page.

Married in India, Pixie was deprived of an orthodox honeymoon, but she never failed to see the humorous side of the picture and accept anything and everything like a better than good sport. She looked forward, starry-eyed, to a romantic adventure, but she found a peripatetic husband always in quest of a fossil bone just over the horizon. She hoped the Vale of Kashmir would lure him to "stay put" for a week or two. Only a few days was all she asked. But did she get them? No. He never even saw the luxurious houseboat she had engaged. She says: "Another unforgettable day I spent at Shalimar, the Abode of Love, amid almond blossoms, sparkling fountains and carpets of brilliant flowers—dreaming my way through marble pavilions where lovely ladies played and Shah Jehan whiled away his summers. This was the very garden of the Kashmiri song Pale hands I loved beside the Shalimar. My hands were pale beside the Shalimar, but who was there to hold them? In all this loveliness there must he love! But my fossil-hunting husband wrote to ask is there any bone? Had I seen anything that looked like a 'skinkus' in the fair Vale? The letter was postmarked 'Chukwal.' I had a notion to fire right back, 'No! but I know where there's a 'skunkus' in a place called Chuck-all!'" So poor Pixie was hustled off to help hunt for a Baluchitherum which eventually my own expedition discovered in the Gobi Desert.

Of course, Barnum found bones wherever he went. He always does. Professor Henry Fairfield Osborn once said to me: "Brown is the most amazing collector I've ever known. He must be able to smell fossils. If he runs a test-trench through an exposure it will be right in the middle of the richest deposit. He never misses."

Barnum has been my friend for well nigh forty years. He is a lone wolf as a fossil hunter. Pixie discovered that to her sorrow. I have known him to disappear from the Museum, just fading out like the Vanishing American, and none of the staff knew where he had gone. It might be India, Burma, Greece, Patagonia, Canada or Wyoming—but invariably his whereabouts was disclosed by a veritable avalanche of fossils descending in carload lots upon the Museum. Barnum himself would follow eventually, just drifting in quietly with no fanfare. One day his office was empty; the next morning there he would be sitting at his desk as though he never had been away.

He has discovered many of the most important and most spectacular specimens in the whole history of paleontology. When he ceases to look for bones on this earth, the Celestial fossil fields may well prepare for a thorough inspection by his all-seeing eyes. He'll arrive in the Other World with a pick, shellac and plaster or else he won't go. You may be sure he'll still be doing it, alive or dead.

But this is not Barnum's book; it's his wife's book from cover to cover. He couldn't have written it; no man could!

It won't be put on my library shelf. I shall keep it on the night table close at hand where, when I am low in spirit, I can always have the rare medicine of its delightful humor. Here's to you, Pixie! Do give us another.

Roy Chapman Andrews

PART ONE

# INDIA

"Oh, Life, I have taken you for my lover,
I rent your veils and found you fair.
If a fault or failing my eyes discover,
I will not see it; it is not there!"
—India Love Lyrics

# 1
## THE BRIDE WORE BREECHES

If you dream of a little place in the country with white, ruffled curtains, cozy fireplace and lilacs in the spring—don't marry a paleontologist.

I did. And overnight my dreamhouse changed into a dreamship, soon to become another Noah's Ark with me the red-haired nursemaid to a menagerie of fossilized freaks.

This transformation really started in a palm-shaded chapel in Calcutta, India, when I married one Barnum Brown of the American Museum of Natural History.

Having finished my schooling in an upstate New York convent, I had embarked with a graduate student group on the first leg of a world tour. The Orient—especially India—loomed large in my imagination. It loomed large also in the plans of a fellow-passenger heading a bone-digging expedition to that charmed part of the globe.

Then as now, Dr. Brown was tall, straight, deliberate and thorough, with twinkling blue eyes that went well with pince-nez. His features suggested the scholar rather than the field-explorer, being a bit on the dignified side. Personally I thought he approached pretty close to the public's idea of what a scientist should look like.

Such was the man who whisked me from altar to jungle for a prehistoric honeymoon. He was a big-game hunter—the kind whose game has been dead some millions of years. In other words, Barnum dug bones for a living, and our so-called honeymoon was largely a bone-digging affair marked by the triumphant unearthing of several ancient monsters.

The first turned up in the Siwalik Hills—a strip of lost world bordering the Himalayas and famed as India's foremost ghouling grounds.

"Boasts more skeletons to the square mile than any other part of the Orient," Barnum had explained. "Bygone haunt of sabre-tooth tigers, and hyena-bears, and extinct pythons, and strange dogs big as lions." Nothing aroused him to a higher emotional pitch than the thought of bone, unless it was actually uncovering some monstrous relic. "God's country," he called that strip of lost world.

To me it looked as if the devil had long since taken title. The region had a bad reputation, too, in a ghostly sort of way. According to local superstition, it was under the curse of Siva, Hindu god of destruction, and the bones were those of ancient giants who had aroused his anger long ago.

Needless to say, we had quite a time getting a camel-caravan together for our enterprise. The natives were taking no chances with ghosts. Those we finally did land were a cutthroaty lot who feared neither the living nor the dead. But we had a good man to keep them in line. Abdul Azziz was his name—a big Punjabi Mussulman from Rawalpindi, with the body of a Hercules and the face of a baby.

We had found him in the home of our dear English friend, Brigadier X, through whose influence Barnum's work in India was greatly facilitated. With his generous wife he lived in true colonial style on a country estate in the Punjab, Abdul his head servant. "A priceless fellow," the Brigadier commented. "Worth 50 of the usual run. Manages my entire establishment."

Yet when it came to parting, our host insisted that we take him with us. His job would be waiting when he returned. The loan of a servant was considered the height of hospitality in British-India, the gesture of a "Pukka sahib"; in the American vernacular, "a real gentleman."

Abdul was everything the Brigadier said—a wonder-working jinni straight out of the Arabian Nights. There was nothing he couldn't do, even the apparently impossible. "Acha, Sahib" (okay) was his motto. You had only to wish, it seemed, and he would appear on the double, eager and ready to transform your slightest whim into reality. Besides cooking he bossed the crew, managed camp, and served as interpreter and contact with the people—the country folk of India. Most of all he was a friend.

Abdul was married and, although Mohammedan Law permitted him four wives, he managed to eke along with three— "one for the heart," as he put it, "and two to pass the time." He hoped some day to acquire a fourth.

"And what would she be for?" I asked.

A beatific smile wreathed his face. "Ah, Memsahib, she—she would be for the glory of Allah."

Our expedition mascot was a Spaniel pup named Taj, alias Lace Pants. A wedding present from my husband, she rode in the saddlebag on the left flank of my horse.

My garb consisted of boots and riding breeches, with blouse and topi, or sun helmet. Barnum usually wore boots and shorts, his shirt open at the neck, and smoke glasses for the sun. A topi on his head, binoculars slung over one shoulder, compass case and clinometer (for measuring dip of rock formations) over the other; camera suspended from the neck down the front; small geologic pick stuck in belt; saddler's awl in back pocket—that was the general picture of Barnum on safari.

Day after day our caravan plodded deeper into the Siwaliks, following the stream courses, thin green ribbons of jungle winding between the hills. Through villages of gaping natives we filed, down canyons whose high walls shut out the sun. Clumps of mango trees, palms and the sacred pipal grew in the sheltered places, forming cool oases where we paused and waited for the heat of noon to pass.

During one of these halts we made our first big discovery. Taj announced the glad tidings. Then Barnum came bounding up, shouting, exultantly,

"I found it! I found it!"

Before I could ask what it was, he had grabbed my arm and was dragging me through the brush to the edge of a dry wash. There he stopped and pointed a long sensitive finger at what looked like the top of a huge boulder protruding from the sand.

"What is it?" I questioned.

"Elephant skull."

I was disappointed. "What's so unusual about that? There are elephants all over India."

Barnum regarded me in silence, then answered, impressively,

"Pixie, this elephant is prehistoric—extinct—millions of years old. The bone has turned to stone."

Abdul panted up with two of the men, wanting to know the reason for all the excitement.

"We've struck pay dirt," my husband answered. "Abdul, have the crew set up camp right here in the canyon. Pitch the tents under those mangoes." He smiled down at me: "How about it?"

"I don't know about your pachyderm," I answered, surveying the landscape, "but you do know how to pick campsites."

My elated spouse pulled a map from his breast pocket and flipped it open on the ground. His pencil circled a small dot marked "Siswan" in the middle of a blank expanse. "That's us," he said. "Siswan village is about a quarter-mile from here. Make a perfect base for working this whole section. What say we knuckle down to some home building right here?"

"It'll be darling!"

Suddenly I felt his lips soft on the back of my neck.

We strolled over to where a place was being cleared for the tents. The camels were down and all hands in a festive mood. The long, hard trip had ended; now for some real living.

Taj caught the spirit too, and skittered about like crazy, dodging between legs and clowning with Abdul. The big man would make a lunge for the little pup, and we'd die laughing. A rhino after a butterfly!

Camels number two and three, first to be unloaded, were the most pampered beasts in the outfit. They carried the "irreplaceables"—sacks of plaster, bottles of shellac, cans of alcohol, packets of rice paper, enough burlap to clothe an army of hoboes. There were little picks, big picks, crooked awls, straight awls, shovels, chisels, cobbler's hammers.

I watched the unloading. One mysterious wooden box stamped with rows of X's caught my eye. I turned to Barnum.

"Dynamite," he said, lighting his pipe. "May have to do some blasting before we're through. When a specimen is buried under several tons of rock, which isn't unusual, it takes more than elbow grease to get him out."

Then came suitcases, a portable typewriter, small boxes of chemicals, cases of something that rattled. Cameras too—three of them; a Graflex for stills, Bell & Howell for movies, and my old box Brownie when all else failed.

"Where are we going to put all this stuff?" I finally asked, beginning to weaken.

"These," quoth My Man, his thumb testing the point of a pick—"these are not STUFF. They are the tools of the highly creditable trade of grave-digging. *My* trade, as you should know by now, little wife." He added, "And they go right over there," nodding towards a large tent. "That's to be our combination supply depot and dark room.

He smiled at my bewilderment. "The dark room's for you, Pixie. Once the work gets under way, there'll be a lot of photographing to do—and you'll have to take care of the developing."

After storing the "irreplaceables" and making the dark room really dark, we took up the question of where to pitch the "home" tents—Barnum's and mine.

"How about under that large mango," I ventured, "and end to end, so we'll have plenty of room?"

Abdul and his gang had them set up in no time and the Browns prepared to move in, bag-'n'-baggage.

Suddenly I gave a half-screech as a pair of long arms swept me off the ground. To the amazement of Abdul and his gang, my Soulmate carried me across the threshhold just as in novels. After all we *were* newlyweds and this was our honeymoon!

Not far distant they erected the cook-tent, plus a lean-to for kitchen extras, while from an overhanging branch swung our "refrigerator"—a large earthen jar encased in thick wet coconut fibre which caught and held each cooling breeze.

The Syce (horsemen) and camel tenders, hardier souls than we, had no need to "make" camp. They ate and slept in the open, curling up in the sand with their horses and camels at the end of the day. Life was that simple.

By nightfall you'd have thought we'd been living there for years. Abdul, our Man Friday, had a fire going and supper on. The three of us stretched out in our camp chairs, Taj in my lap, hubby smoking his pipe and thinking about his "find." Out on the river wash dusk gathered, deep shadows closing in from the sheer sides of the ravine.

All at once a strange sensation came over me. I felt definitely uneasy. There was a curious prickling along my spine.

"Are you sure this is a—a good place to camp?" I finally asked, in a low voice. "I mean are you sure it isn't on somebody's property, or something?"

My husband cupped his left hand over mine. With the other he massaged behind his right ear, a Barnum Brown gesture indicating held-in laughter.

"Don't tell me you believe the ghost stories they've drummed up about these hills," he said. "Probably just a few natives from the village out to give us the once-over."

But the feeling persisted; the feeling of unseen eyes staring out from the shadows. The black growth beyond the tents seemed alive with invisible things. The air took on an eerie stillness like the calm before the storm.

"It gives me the creeps," I declared, and remember adding, just as we fell asleep, "I have a wonderful name for this place—Spooky Hollow."

## 2
### SPOOKY HOLLOW

The attack came at dawn, shattering my slumbers with awful screechings, yammerings and chatterings. The tent shook under a rain of missiles that strummed against the canvas. At the risk of my life, I took a quick peek through the flaps.

An amazing sight! All around were flocks of monkeys. Monkeys swinging from trees, scrambling over rocks, leaping, fighting. Crabbed old patriarchs vied with agile youngsters. Angry mothers carried their babies underslung-fashion. One group was hopping madly up and down like rooters at a football game.

Then I saw the reason. We were being raided. Five hairy brown simians were hot-footing it down the gully with odds and ends of camp equipment. One was struggling to carry off several of our aluminum cook-pots; another trailed boots and breeches after him. The remainder, armed with soup ladle, pick, and Abdul's hookah pipe, strongly suggested The Three Musketeers.

Barnum, his face still lathered, and only half-dressed, was tailing the monk with the pants, and yelling, "Come back, you dirty brute. Come back!" The last I saw of him was a flying shirt-tail disappearing in the brush. Seconds later Abdul streaked past brandishing a fistful of kitchen cutlery.

The commotion had our livestock in a dither. Taj barked bloody murder. With horses and camels about to take off, the men had all they could do to hold them down.

Hurriedly I slipped into some clothes and started after my husband. But no sooner had I popped my head out of the tent than—

zingo!—something missed me by an inch. Snipers covering the retreat!

Undaunted, I advanced a short distance into the open, only to be stopped dead in my tracks by a hail of pebbles and sticks. The beasts were laying down a barrage from some nearby trees. I fired back, but most of my shots went wild. Just as I would take aim Mr. Monk would swing down by his tail and seem to thumb his nose at me upside-down. He made a darned poor target.

Then an idea struck me. How about using a little psychological warfare on them—something that would scare them off? That big tin basin in the cook-tent was just the thing. I dashed over, hauled it out, and with an iron spoon stirred up the loudest racket I could. The ravine filled with the booming which rolled on, echoing from wall to wall to the farthest crag. In a twinkling our tormentors were gone.

It saved the day for Barnum, too. "Thank God for that noise," he remarked on his return. "What was it—an avalanche?"

"Just me and this plaster-pan," I said.

My husband regarded me proudly, then stooped and kissed me. "You've earned the highest decoration available at the moment," he said. "When those thieving imps heard the banging, they dropped their loot and ran."

Breakfast was a hurried affair that morning, and none too good. One of the monks had smashed Abdul's hubble-bubble, the coffee was bad, and with the confusion of getting settled things were in a mess.

Pants retrieved all in one piece, the Hunter couldn't wait to hie himself off to work. Fortunately, the elephant pit was practically in our front yard, so I was able to watch operations from camp. Each time I glanced over, Barnum had sunk a little deeper in the hole. Taj became an on-the-spot reporter, emitting shrill barks of alarm whenever Barnum slipped in altogether. Then she would tear back to check on me in her efforts to keep track of both of us. On the tenth or twelfth trip, however, she crawled under my cot and made herself scarce the rest of the day. I learned later that she had misbehaved on the back of the skull— "right on the *occiput*," the scientist complained.

There was still much unpacking to do, and interior decorating. But, somehow, I couldn't put my mind to it. It seemed more important to keep Barnum supplied with hot tea at the dig where I could better observe the unearthing of our titan. The work was slow and tedious. By inches, Barnum's awl and soft brush uncovered more and more of the huge skull. As fresh surfaces of bone were exposed, he gently brushed quantities of shellac into them, the syrupy liquid soaking in and drying hard as rock.

When not plying my husband with tea and foolish questions, I would make another stab at helping our head-servant with the chores. But now distraction came from another quarter—the jungle—and from time to time I found myself wandering off on small exploring expeditions all my own. On one of these, *I* made a discovery. Till now, everyone had been so busy that our surroundings had received scant attention, so I was sure to be alone in my secret.

Secrets will out, however, and this one did at the evening meal. "Would you be interested in knowing whom we have for a next-door neighbor?" I remarked, temptingly.

Barnum's brows arched. "Didn't know we had one," he said, adding, "surely not those hell-raising monkeys?" The brows knotted in a frown.

"There's a Hindu god living not a stone's throw from camp."

The scowl cleared. His head tilted quizzically. "Which god? There are thousands, you know."

"Ganesh, the Elephant God of India, no less," I gurgled triumphantly. "Care to meet him?"

Rising, I led the way through the jungly growth toward a high rock wall covered with vines. My husband followed, good-naturedly, thinking it all a huge joke. Once over the wall, I stepped to one side and dramatically pointed out the cute Hindu temple I had found earlier in the day.

Moss-grown, almost hidden in vegetation, and now in the dusk looking more ancient than ever, was a small masonry octagon rising to a full rounded dome. Before the pillar'd portico at its entrance, on a raised platform, stood the stone image of an elephant—the sign that this was a shrine to the God Ganesh.

"By golly, you were right," was all he said. But his smile was broad for Ganesh is the god of good luck.

That night we prepared for further monkeyshines. I placed the plaster-pan and iron spoon beside my cot. Barnum put the cameras and mapcases beside his. Abdul gathered a great pile of dried brush and set it to one side, assuring us that there was nothing like a good powerful blaze to put the jungle cut-ups in their place. Then we checked on the tent pegs to make sure they were firmly anchored, tucked all moveable equipment in hard-to-find corners, and set a booby trap—a low wire strung down the middle of camp, one end attached to a precariously balanced stack of pots and pans in the kitchen-tent.

"Hatches down, Cap'n. Decks cleared for action!" I reported as we turned in.

Barnum massaged behind his ear before replying, "All we need now are some barbed wire entanglements and a couple of pill-boxes."

Morning found the monkeys back—but as friends, not foes; and I soon learned that nothing was safe from their busy fingers, of which they had twenty instead of ten. Their mania was soap and bathroom accessories; the pride of one young buck, my hotwater bag. So proud was he of his new possession that he wore it on every occasion—slung over his shoulder for a knapsack, on his head for a cocked hat, or in his arms for a ukulele. Every time "Sportin' Life" traipsed by with the bag, the monkey maidens fairly swooned. One of them eventually hooked him, and I watched the pair making love on a nearby slope, picking each other's posies and munching berries. But even at the height of his passion Lover clung to his decoration. Apparently he believed that sex, while a lot of fun, wasn't everything.

While I spied on the spooners, plenty was going on behind my back. A sprightly trouper was doing a Dance of the Seven Veils with a roll of toilet paper; two others, climbing into a provision box, had jiggled the lid closed and now couldn't get out. It was one of the funniest performances I have ever seen—though it evoked nary a smile from Abdul.

Much as he loved animals, Abdul had no use for monkeys. "Baby-snatchers," he called them. "They'll bear watching, Mem-sahib. Many times have I seen them drop from trees at the edge of a village and carry off some little child when its mother wasn't look-ing." The big man's face softened a trifle. "Oh, they treat an adopted baby gently enough for a while, cuddling and fussing over it like one of their own. But later, in an off moment, baby is set out on a limb to dry and falls through. Some villages have old men with special 'monkey calls' who coax the baby away from Mama Monk before it is too late."

"In lieu of a baby, maybe they're planning to make off with me," I laughed. "They're friendly enough."

"Friendly's the word," Barnum snorted. "Looks as though we've been accepted into Spooky Hollow society."

A complete change had come over our little world. The valley had suddenly sprung to life, its emptiness filled with the sounds of wild things unafraid. Songbirds fluttered in the willowy neems, and the very leaves danced to the singing. Lizards scuttled over gleam-ing boulders. Small, friendly rustlings occurred in the underbrush. Overnight we had become part of the life of Spooky Hollow. We belonged. And the ravine walls appeared to close in, cuddling the valley between them, making it cozy and homelike.

Each morning, like emissaries from some fairy court deep in the glades, came peacocks, trailing beauty to our table and request-ing their tribute. Residence in this their valley had a price. But unlike the uncouth simians who took their rent out in plunder, these gentle creatures asked only tidbits from the table—crumbs.

Between crumbs they entertained us with a floor show. Against the dark backdrop of wild ravine with the first rays of dawn filter-ing through the trees, the bird ballet swirled 'round our breakfast circle. No costumer could have matched the brilliance of their glis-tening plumage. No Billy Rose could have staged a more perfect setting. No Pavlova could have danced with more exquisite grace. Through the mottled sunbeams they moved, the bright bold males strutting and preening as, with feathers ruffled and fan-tails flar-ing, they slowly pivoted like models at a fashion show. Around them tripped the smaller excited hens. Off in the wings, a flock of fluttery quail put on a show of their own.

There was music, too, from a thousand tiny bird throats—thrushes, sparrows. We could hear the solo of a koel, the Indian cuckoo, singing in the topmost branch; the soft cheery notes of the bulbul chorus; the whistling of the drongo.

In time, all the birds in the valley came to the Browns for their daily song and dance. Some stayed to lunch and dinner, and these grew so fat and sleepy they couldn't carry a tune anymore, except for an occasional "beep."

## 3
### Curry and Bones

Time, which slips by swiftly when one is happy, fairly raced those first glorious days in camp. And they were busy days. Barnum was almost up to his neck in bone now. He had excavated a deep trench around the skull, and all I could see were his head and shoulders above the stream-bed. A few yards to one side he had located the lower jaws—huge things with wide grooved teeth. Nothing of the skeleton had as yet come to light, but he was having enough worries as it was. His current headache was the long tusk, just uncovered.

"Don't know whether I can save it," he remarked dolefully. "Old ivory may be okay in an antique shop; it's a nuisance on fossil elephants. Begins to break up as soon as the air strikes it." He let more shellac seep into the chalky under-surface, deftly pressing thin sheets of rice paper over the sticky coating before it dried. "Keeps the small splinters in place—or should. Chances are it'll continue crumbling just the same. Never knew a tusk to hold together—ever."

As for my headaches, they were shared by Abdul. I don't know what we would have done without our dusky giant, and as the days passed I was learning to appreciate him more and more.

A close bond of affection had sprung up between the big Moslem and Taj. They thought the world of each other and neither was completely happy when the other wasn't in the offing. Only one thing threatened the friendship—Pup's disregard for the Mohammedan religion.

23

Five times a day, as required by his faith, Abdul would salaam in prayer toward Mecca. The morning, noon and afternoon observances usually passed without incident, but those at sunset and nightfall were different. Taj felt frisky in the cool of the evening, and when her giant companion unrolled his prayer rug and got down on hands and knees, she thought he was playing dog.

Head down, tail in air, she would start in with a nip at his heels, dash around front to nuzzle an ear or lick his nose, then double back through the middle between his legs. To all this the devout Moslem paid no heed.

But Abdul wore huge baggy pantaloons with plenty of room inside for a little dog to go adventuring, and while her man was down on all fours Taj had easy access to the interior. Whenever she vanished during vespers it was only a matter of seconds before she reappeared as a wriggling lump in Abdul's pants.

Up one leg, down the other, the lump would go, emitting small muffled yips from time to time. Sometimes, however, in crossing from one limb to the other, the pup became caught. Then claustrophobia seized her and, in panic, she threw the pants into an uproar. Abdul was ticklish, and, try as he might to keep a straight face, his prayers would end in peals of laughter.

"It's the devil sends this imp to torment me," he'd say. But, between playing and praying, I think Allah enjoyed a good laugh, too.

I liked working with Abdul. Everything seemed easy when he was helping you. And while Barnum fussed and fumed and stood on his head trying to raise the dead, Abdul and I were busy transforming the wilderness.

With a bit of doing, camping in India can be a most luxurious affair. Don't think we were forced to rough it because we lived in tents. True, we lacked much of the padding of modern living, but we had the essentials, and all-in-all were quite as comfortable as in our own home.

Servants? You can hire a dozen for the price of one in America. But then, it takes a dozen to do the work of one in America, so matters come out about even. The reason for this is that there are no jacks-of-all-trades in the country. They are prohibited by the caste code which is as hard to crack as one of our Union laws.

Each domestic is a specialist—born to the broom, the water-pot, the cookstove, or whatever. To perform any other labor than that inherited at birth is strictly taboo. For instance, the *bhisti* (water-carrier) will have nothing to do with washing or cooking and wouldn't be caught dead hauling anything but water.

Water-carrying being of vital importance in India, *bhistis* are above average, as servants go. Their devotion to duty has often been noteworthy, as in the case of one awarded the Victoria Cross for getting water through to a stranded regiment under fire. Such a *bhisti* inspired Kipling's "Gunga Din."

Now, the *dhobie* is different. His trade is washing, when he works at it—usually under threat. His talents lie much more in the loafing line. But it's just as well that way; next to moths, silverfish and fire, the *dhobie* is public enemy number one, speaking laundry-wise. He is an expert at beating the clothes out of dirt and breaking a stone with a shirt.

*Khansamahs* are not so bad. They are the cooks. Our *khansamah*, Fasil, was really more of a chief bottle-washer, preparing the meals only when Abdul's duties as head-servant required him elsewhere.

Between them all, there were few dull moments. While trying to figure out who couldn't do what and why, things somehow muddled through.

Even Abdul was a problem sometimes. We had our differences— not often, but often enough to keep the relationship from bogging. Our first tiff was touched off by a combination of bacon and Mohammedan—a highly combustible mixture in the Orient.

He and I were storing some foodstuffs in the provision tent. Taj, as always, was attempting to make a picnic of it and succeeding, when, all at once, my partner commenced beating his forehead and mumbling in Urdu, his native tongue. He peered dejectedly into a large sack, just opened, obviously distressed by what he saw.

"Why the grief?" I asked, cheerily.

"The sugar is spoiled, Memsahib."

Something hit me in the pit of my stomach. Sugar was a scarce article out here in the sticks. "What are you saying, man? That sugar was especially packed to keep dry. How could it go bad?"

Abdul threw up his hands in despair. "Someone stupid. They packed the sugar in the same sack with the bacon."

"So-o-o. Is that bad?"

"Bad?" he boomed. "Bismillah! In the Name of Allah! Bacon is pork. Pork is unclean meat. Everything it touches is spoiled. The sugar is worthless now. A good Mohammedan wouldn't touch it."

We settled the matter by exchanging his sugar for our own private stock which had been packed with the canned goods. As Christians, and therefore beyond all salvation, it didn't make any difference what we ate.

Thereafter Abdul had the handling of all food. He would have it no other way. I was only too glad to agree—what with his abhorrence of pork as a Moslem, with the anti-beef policy of our Hindu cameleers and horse boys, and with sundry other differences in the diet. Arranging the menu for a mixed crew of Indians is just another of those Oriental riddles the Westerner can never hope to solve.

But Abdul was really a wizard when it came to cooking. There was hardly a thing he couldn't make delicious. Nor did he need a magic lamp to do it; any old pot or pan would serve. A little water, some rice, a dash of spice, meat, a few charcoals and, with a puff of smoke—Allakazam! Dinner! It wasn't cooking; it was alchemy.

An Aladdin's Cave of epicurean treasure, the cook-tent became my favorite spot in camp; and Taj's also. Both of us would stand and sniff the fragrance of that wonderful place by the hour—the hot, fresh-ground spices, the ginger, coriander, chilli, turmeric—and grow hungrier and hungrier.

Stacked high in the rear were American canned and British tinned goods (if you can figure out the difference) with vegetables, potted meats, bacon, ham, butter, fruits, preserves and good coffee. Fresh meat, too: mutton, goat, beef; and Abdul always had a few live chickens around fattening up. Plenty of sesame oil, of course, and rice, fresh eggs, and dhal—dhal—and more dhal.

Dhal is the all-purpose food of India, a native grain much like our dried peas. It can be prepared in any number of ways—mashed, hashed, boiled in soup or cooked in curries. What it tastes like after all the fixing is anybody's guess.

Milk was another thing we had plenty of, especially buffalo milk. The local *dudh-wallah* (dairyman) kept a herd of buffalo up the draw, and delivered two seer (four quarts) of fresh milk to our door every morning. It was really more like heavy cream, Abdul using it in many of his dishes—soups, sauces, puddings and curd sweets. Naturally, we boiled some for drinking, also the goat's milk. This boiling procedure I saw to myself, for there exists a wide difference of opinion between East and West as to what constitutes the boiling point.

It was fascinating to watch the big man go about his work. With an eye to picking up some ideas on Oriental cookery, I often pitched in and helped, especially when he was making curry. What made his curry so unctious was a deep secret at first.

"Simple," said Abdul, trying to put me off. "You just take two onions—good size and chop them up in three tablespoonsful of oil. Curry powder next, one or two teaspoons—to taste, mind you. Salt. Then add the meat—two cups, raw or cooked—whatever is handy. Fry very slowly; too much heat no good." He bent to wave the smoke away. "And there you have the curry," he added with an air of finality. But I could see that he was laughing.

"That's ordinary curry, Abdul. But the secret of yours—the secret," I teased.

"Well, only for you, Maharane, I tell all." The man did a little dance over to the shelf, broke open a coconut and, with a flourish, poured a teacup of juice into the curry and left it to simmer uncovered. "That's it, Memsahib. That's the secret!"

"And the touch divine," I applauded. "Curry 'n rice for dinner, Chef?"

He shook his head. It was curry 'n *noodles*, the secret of which anyone could smell. "Two cups of curry to four of boiled noodles are the proportions," cook went on, gently stirring the mixture over a slow fire. Immediately before serving he mixed in a diced raw onion and dill pickle, topping the dish with a thick crust of fried noodles, crisp and brown. I gave an imitation bark, at which Taj licked my chops. "Indian Noodles" was the specialty of the house.

Then there was chutney. No one ever visits India without meeting someone who makes the best chutney in the world. And that is as it should be. Chutney is an original Indian condiment, albeit in the last century or so it has acquired a somewhat British flavor. Due to a little experimental blending on the part of Major Gray and other epicures, the chutney of today is quite as British as mangoes, ginger and red chillis will permit. It is the standard sauce for curries, meats and such on all proper English tables throughout the Commonwealth.

In the condiment line, Abdul's talents all leaned toward chutney—and with reason. His former employer, being a British Brigadier, was an inveterate chutney-eater and Abdul had had to learn the art of making it, or else. He did, with flying colors, coming up with a variation that shamed all the other *bobberjees* (army cooks) in the regiment and was soon the pride of the officers' mess.

The trick was done by taking three pounds of mangoes (peaches will do) and slicing them up fine with an equal amount of sugar, this kept at a slow boil till "jammy." Garlic and ginger, five ounces of each, were then cut up and in they went, followed by an ounce—more or less—of crushed red chillis, a pound of raisins, pint of vinegar and two ounces of salt—plus plenty of stirring. Simmered down to a thickness, the result was bottled or jarred when cool. Nothing ever tasted better with roast lamb, chicken or pork. Try it!

Needless to say, the Sahib too was pretty much in evidence along about mealtime. Nothing short of a major disaster or a broken bone could have held him back. Most days the smell of the cooking fetched him long beforehand and he would snoop around, peeking under pot lids, spoiling his appetite with dibs and dabs, or stand rooted before some steaming kettle, hypnotized by the rich bubbling within.

When the wind wasn't in the right direction, Taj served as dinner bell. "Go get the papa," I'd order, and off she'd scurry, filling the valley with her call.

There was something festive about dinner under the mangoes. Always it was an occasion. I don't know which was greater—our delight, or Abdul's satisfaction. It was uncanny how he would suddenly appear at the end of a long hot day, looking fresh as a babe in immaculate white pantaloons, jacket, pristine turban and flaunt a full five-course meal as if he had called it down from the heavens. And he knew the art of serving it in the grand manner—the full tray held high, balanced on fingertips, the lofty bearing, the sweeping gesture.

Only once was the delicate harmony of that scene shattered. A dive-bombing crow mistook the curry dish for home-base and sent the dinner cascading down our servant's front. A pick-up supper with self-service did for the Browns that night. Next day Taj had her first lesson in the extermination of crows.

Often after the evening meal Barnum and I strolled up the ravine to where the camel men squatted around small fires baking their *chupatti* cakes, the dry unleavened "tortillas" of India. Or we just relaxed in our comfy chairs, talking and watching the smoke from the dung fires mingle with the dark as day ebbed away on the other side of the rimrock and night rushed in, cool and quieting.

Our peacock friends gathered 'round for one last twirl before going to roost, the feathered small fry cheeped sleepily in the branches overhead, and from somewhere near the temple came the determined bustling of monkeys putting their beds in order—sounds of Spooky Hollow tucking itself away for the night.

It was then, with my family there beside me, that I knew this was the happiest time of my life.

## 4
## VISITOR BY NIGHT

One evening Barnum and I climbed a nearby hill to watch the sun set in crimson glory. The rocks were molten gold, and in the valley below we saw a water-hole where lazy buffalo wallowed in the mud. It was a strangely silent twilight hour. I suddenly realized that the chattering of the day had ceased; that the monkeys had disappeared. On our return to camp I questioned Abdul about it.

"Cats," he explained, glancing furtively about. "Monkeys always leave when cats come."

I had visions of the alley variety and was about to query him further when some duty claimed my attention and the matter dropped.

Came night. A full moon silvered the river wash; threw ghostly shadows among the cliffs. Our camp slumbered at last. But the beauty of the jungle-world kept me awake. I lay listening to its weird noises. Mysterious howls and tremulous cries filled the valley. An owl hooted the hours—twelve—one. At two o'clock the only thing stirring was a cup of coffee that I poured from my thermos before falling asleep.

Next—I was sitting straight up in bed, every nerve tingling, ears straining for some elusive sound. Dead stillness. Whatever it was had awakened Pups too, as, shaking and whimpering, she came crawling toward the head of the cot. I gathered the quivering form in my arms and tried to quiet her, but the trembling only increased and her breathing came in short quick gasps. Something was out there!

Slowly, carefully, I moved to the front of the tent, drew aside the flap and peered out. Moonlit sand. Black shadows. I watched and waited, feeling Taj's heart pounding madly against me and I, too, began to tremble.

Presently, one of the shadows moved. It detached itself from the others. Then into the brightness stepped—a leopard! That was Abdul's cat! Dark and menacing, he loomed against the sand, yellow eyes glowing. He stood there a moment like a statue, terrifying, yet beautiful. Then he melted back into the night. Only the glint of his eyes could I see now, moving first this way, then that. When they seemed to brighten and enlarge, I knew that the beast was creeping forward. No sound beneath the footfall, no sign of body—just those two burning orbs drawing closer—closer.

I followed them—hypnotic eyes that held my gaze so that I could not look away. I wanted to turn and awaken Barnum; my limbs were paralyzed. I tried to cry out, but the sound was muted by a terrible dryness in my throat. There were only the eyes and the blackness—and then only the blackness—

A voice asked, "Since when have you taken to sleeping on the floor?"

I opened my eyes. Daylight! Barnum leaning over me, a bottle of smelling salts in one hand!

"I—I guess I passed out," I offered, apologetically. "The leopard—what happened to it?"

My husband looked surprised. Then he laughed. "Leopard? What are you talking about? There wasn't any leopard here, Pixie. You were walking in your sleep and fell. This came down with you. Look!" He picked up his metal canteen from the floor and hung it back on its hook. "Must have been a honey of a nightmare," he concluded, still smiling.

Nightmare! Some time passed before I could think of a comeback. Nothing adequate occurred to me until we were leaving for breakfast—when the answer lay glaring dramatically up at us from the sand. "It's the first nightmare ever to leave a trail behind," I announced, triumphantly, indicating the fresh pug marks that circled the tent.

They were leopard, all right. Barnum admitted as much after examining them. "Saved by a canteen," was his soothing comment. "Lucky you knocked it down when you fainted. Undoubtedly the noise was what scared him away.

"Loud noises do seem to come in handy now and then," I observed. "They saved your pants from the monkeys, my skin from a leopard. The wildlife hereabouts evidently prefer white meat."

Abdul cut in. "The leopard was not after you, Memsahib. He wanted Taj."

"Taj?" My heart skipped a beat or two.

"Yes, Memsahib. Dog is the leopard's favorite food. Look, I show you."

A large wooden box lay concealed in the brush not far from camp. Fearfully and wonderfully made, it had heavy plank flooring, sides and roof, plus an iron-barred door at one end that automatically let down and locked when anything entered. According to Abdul, this was the Siswan community leopard trap—but as it never killed anything, even a mouse, apparently it had been built by a Hindu.

Inside, closed off in a protected compartment, was the bait—a poor little flea-bitten pariah dog who didn't enjoy his role a bit. Nor did he look like much of a meal.

Obviously, the local felines were of the same opinion. The earth surrounding the trap was covered with their tracks, each morning fresh ones; but never a sign of an animal having crossed the fatal threshold. Those leopards were just too smart.

And while we waited vainly for one with more appetite than brains, I was kept busy feeding the hungry bait.

5

## Hizzonor the Mayor

News got around fast in Siswan. It wasn't long before our "goings-on" in the gully had the natives wondering, then investigating. They started on me.

I had dropped off to sleep under the mangoes of an afternoon, awaking to find myself surrounded by a mixed delegation of Siswan citizenry. At first I thought it must be the monkeys; but monkeys didn't wear clothes, nor were they in the habit of ogling one with such undisguised curiosity. Under the circumstances I was hardly in a position to "entertain," so I just sat up and smiled, friendly-like.

That did it! The crowd broke ranks and swarmed around me, jabbering, laughing. Peace and privacy flew out the window, and with them went any hope of honeymoon seclusion—even make-believe. From that moment forward, camp was "Coney Island" for every man, woman and child within walking distance. They came in groups and stayed all day, their one object to watch the white sahib and memsahib. Barnum and I seldom had a minute to ourselves.

Not wishing to miss anything, the audience arrived well before dawn, and would be waiting expectantly when "Barrymore Brown" and his leading lady poked their noses outside the tent. Front rows were always filled, the animal folk well represented by goats, dogs and a mongoose or two. The balcony was packed with hoi-polloi peering over the temple wall.

Our appearance brought a hush. "Your public, Madame," the hero would whisper, holding aside the tent flaps. "Strike up the band, maestro. On with the show!"

Our first bow to the gallery always elicited squeals of delight. The comb-and-brush act stirred a round of applause. The tooth-brush skit brought down the house. Every move was followed with rapt attention. Even the lacing of my boots struck someone's funny bone. Without a doubt, the show was a howling success, though I think we overplayed our parts a trifle for an extra laugh. Laughter is such a rare commodity in India.

When the party really warmed up, it reminded us of a circus. All we needed was soda pop, sawdust, and the conventional barker out front. "Hur-ry! Hur-ry! See the funny people from America. They walk. They talk. But are they *human?* See them eat, see them sleep; see the lady that looks like a man. See the death-defying act with knife and fork."

This last, performed at every meal, evoked the most profound wonderment. It involved no more than the ordinary use of these articles, yet never failed to bring the people to their feet yelling and perilously craning their necks for a glimpse of the plates and the objects of our destruction. The fork especially intrigued them. Advocates of the finger method themselves, they couldn't under-stand how we managed to survive so many repeated jabbings in the mouth.

In time we grew quite attached to our fans, Barnum avowing that camp wouldn't be the same without them. Oddly enough, they were never in the way, nor can I recall them giving us one unpleas-ant moment. Two in particular became more or less permanent figures in our camp life.

The first of these was an odd-looking, middle-aged character whom Barnum discovered one day at the edge of the jungle, a little native man standing stiffly erect, patiently awaiting our attention.

Except for a gay-colored turban, he wore European clothes. They weren't exactly Bond Street, but then what did tailoring mat-ter in a river wash in the middle of India? The material was duck—white duck once—and though the pants were baggy, the coat fit like a grape-skin with cuffs stopping short halfway up the arm. A white shirt collar, buttoned without benefit of tie, encased his neck; flashy, yellow leather shoes did the same for his feet. A large black umbrella added the final touch.

"Some punkins," my husband observed. "Let's see what he wants."

When we had approached to within talking distance, the stranger unbent at the waist and bowed low, pressing the palms of his hands together and touching them to his forehead.

"Ram—Ram," he said.

"Ram—Ram," Barnum repeated after him, returning the greeting.

For lack of anything better to do, I smiled and added a "Ram—Ram" of my own. Like the Muslims' "Salaam," "Raming" is the first step in the Hindu ritual of saying hello.

Shown to a camp chair, our guest sat for a long time viewing the premises in stony silence.

At length, jamming the point of his umbrella into the ground, he announced, "I am Tika Lal—*Tehsildar* of the village of Siswan" . . . the words spoken in a monotone—jerkily—yet each snipped off clear and precise as if with a wire-cutter.

"What's '*Tehsildar*'?" I whispered to Barnum.

He shushed me, and answered, out of the corner of his mouth, "The Mayor!"

Our expressions of delight at his visit thawed the man's reserve. His broad grin revealed a set of betel-stained teeth—beautifully matched, both of them.

Becoming suddenly voluble, His Honor launched forth into the traditional how-were-we's. How was our house? he inquired first—this being the stock conversation-opener in India, equivalent to asking how the wife is in America. Another man's wife is never mentioned by a Hindu, even to asking after her welfare. He might be thought to have designs on her. From our house the subject drifted to our health, and from there it just sort of meandered. How long had we been in India? How long did we intend to stay? How was the King of the United States? Where were we bound next? . . .

None of which was meant to be answered. It was simply his way of beating around the bush before switching the topic to himself.

He would have called on us sooner, he explained, had it not been for the press of business. His position as one of Siswan's principal cotton manufacturers permitted him, at this particular season, but little time for his duties as Mayor and official "greeter." Things were going along so he could get away now. Yes, the cotton market was good—thanks to the *Mahatma*. And his brother in Poona was down with the cholera.

As he talked, the Mayor "chewed"—checking his monologue at unexpected moments to shower the earth with reddish-brown betel juice. Taj, I noticed, kept well out of range under my chair, following, with sharp anxiety, the trajectory of each new fall. A shift in the wind brought one dangerously close, upon which she jumped into my lap and would not be put down.

When the quid in the mouth was exhausted, the man produced an ornamented brass caddy from his pocket, removed the lid and a tiny inset tray and placed them on the chair arm. From the lower compartment of the box he withdrew a betel leaf and in its center dabbed a smear of white lime taken from still another small box retrieved from his other pocket. From the tray he took a betel nut and a brass cracker, crushing the nut and meticulously arranging it on the leaf along with cloves, nutmeg and a dash of other spices kept in the tray. These he rolled up carefully in said leaf, turned it over admiringly in his hand, and, with a deft flick of the forefinger, popped it into his mouth. Being a man of means, Tika Lal also used tobacco paste in his "pan" chew, as it is called, and, on rare occasions, possibly a pinch of opium.

The use of betel in India is a national habit like our gum chewing or cigarette smoking. Everyone, high and low, indulges. As a result, the growing and marketing of betel is a highly selective profession. Its harvesting is a matter of special attention and must be performed only by those spiritually fit to do so. There is a legend that should an unworthy person so much as touch one of the plants, it will immediately die. In the bazaars throughout the country one finds "pan" shops—small open booths stacked with the moist green betel leaves kept constantly wet and placed in the shade; tobacco cakes; bowls of white and colored lime, and all the other ingredients which so delight the Indian's taste and make for his greatest dissipation. He will go hungry, rather than do without his "pan."

"Of course, the lime rots the teeth," the *Tehsildar* admitted. "But what is that compared with the pleasure it gives! And there are always those who say it is good for the stomach."

Finally the conversation eased around to the purpose of his visit—why were we camping there? He heard that it was to dig up some old bones, but had dismissed the rumor as an idiot's tale. Who would travel all the way from America just to dig bones in an Indian gully?

"But, that's exactly why we're here," Barnum assured him. "Didn't the Resident Commissioner inform you of our plans to prospect this area? He must have told you that we might make camp somewhere near."

The man's jaw dropped. "A courier did bring information. Still, I couldn't believe . . ."

"Aren't there legends among your people of strange monsters that once inhabited these hills?" my husband cut in.

"It is true, Sahib. There are many stories. But I cannot understand what anyone would want with the bones of these old animals—except to make medicine, maybe."

"They're highly important to us," Barnum explained, patiently. "They tell what the world was like in the distant past, and add to our knowledge of the earth in a lot of ways. We've located a skull in the stream bed here already and are out to find as many more specimens as possible." He eyed the Hindu speculatively a moment, then added, "We'd very much like your cooperation, Mr. Lal."

"You have it, Sahib," came the spontaneous reply. "I shall do all I can to be of service. Have you any immediate needs?"

My husband hesitated; reached over and tweeked Taj's cold, rubbery nose. He seemed amused. "Ye-e-s," he ventured, half in jest. "We'd like you to get rid of a leopard for us. Scared my wife nearly to death the other night. That trap of yours up the draw isn't worth a hang."

To my surprise the small brown hand gave the umbrella handle an angry wrench. His voice rang out clear and defiant: "That I will not do! To kill is against all that is sacred to us. We are Hindus. That the trap does not work is no concern of ours."

The *Tehsildar* reflected before continuing, "Why does not the Sahib destroy the beast? It is not against your faith. Has not the Sahib a gun?"

Barnum confessed that he never carried a weapon of any kind, and, as for killing—except in bagging a good specimen—he was as much against it as any man.

At this, the dusky face beamed again with good will. The Mayor seemed very pleased. Rising, he wished us success and departed with an earnest "May I come again?"

We had made a friend.

Tika Lal was around so much he was practically living with us. Not that we minded. He was a friendly fellow, spoke fair English, and proved very helpful in our dealings with the natives. When he wasn't in camp, you found him in town, defending his title as Mayor or running his cotton mill. According to him, he was something of a whiz in a business way, and I had been promising myself for a long time a trip in to see what Siswan considered a big-time cotton operator.

Tika was the only one of our native guests who showed the slightest glimmer of what the big hole in the river wash was all about. At least he registered something—even if it was perplexity. The others would just bat their eyes in utter incomprehension; the old ones sitting for hours at the edge of the pit muttering to themselves; the young bloods staring blankly at nothing, like so many blocks of stone.

6

## Singing Milkmaid

Our other particular friend was the *dudh-wallah's* daughter, a sweet young thing who answered to the name of Bulbul. We called her that because she was forever singing—just like the nightingale-bulbuls. Her real name was unpronounceable.

If all Indian shepherdesses were like Bulbul, it's easy to see why, according to legend, the great Lord Krishna found them so enchanting. How Bulbul had remained so long unmarried in a country like India I could never understand; and when I'd chide her about it, she would only laugh and hum another ditty. Though not exactly "gay" in our sense of the word—even the young Indians have a dignity and restraint beyond their years—she was a cheery soul, with that deep unaffected sincerity so common in her kind.

We had become acquainted first when her father gave her the task of delivering our milk. Each day the girl would remain a little longer in camp until finally she was staying all day—pottering around, helping with the chores, and brightening things up in a wonderful way. For her, camp was an amazing new world—a place to rummage through tents, handling the endless array of mysterious gadgets and strange "white man's magics" and, strangest of all, the folding cots and collapsible chairs. Curiosity led to quick understanding, with Bulbul shortly appointing herself my personal *ayah* (maid).

As such, she seldom permitted me out of her sight. Wherever I went, the farmer's daughter tagged along, Taj yipping at her heels. On many of our rambles, I noticed quite a tomboy streak in her . . .

the way she'd race the pup, climb trees, scramble over walls. It was all most undignified for an Indian young lady.

But Bulbul cared nothing for dignity. She was free, free as the birds; a little pagan, natural as the day she was born. Never a thought did she give to anything but living. Neither past nor future existed for her. Marriage, caste, religion—those many things that made life such a serious, even tragic, affair for most Hindu maidens, had no meaning for her. The reason I was soon to learn when visiting her people in the village.

Bulbul's hobby, if she could be said to have one, was collecting tinfoil. Of all our possessions, tinfoil was the thing she coveted most, and, as many of our supplies came wrapped in the material, she was always plentifully supplied. What could a milkmaid want with tinfoil? One evening, as she was about to go home with another load, I asked her.

Her eyes twinkled. "Memsahib will see," she said.

I did see. A few days later she appeared in camp with two girl companions, a long red homespun *sari* draped between them. They brought it directly to my tent and spread it out before me.

"We making it for you, Memsahib—a holiday *sari* like those we wear for the festivals," squeaked Bulbul exultantly.

There was the tinfoil covering the garment in a mosaic of bright sequins, each fleck of metal fastened to the cloth with resin and forming, in places, intricate sunburst designs, whorls and circles. The girls were overjoyed at my delight. It was most sincere. A gown from Saks Fifth Avenue couldn't have pleased me more.

"Look," I called, rushing over to show it to Barnum.

He eyed me around a corner of skull, gaped in amazement, then burst out, "Where in the world did you get that?" Followed by "Well, I'll be—" when told.

Exchanging gifts in India is something of a problem, particularly when the other party is Hindu, like Bulbul. She herself had small care in the matter. But there were so many things her parents would have disapproved of on religious grounds, the strongest taboo being on our food or any articles which might later be used in their cooking.

"How about money?" my husband suggested. "Never knew a girl with an aversion for that!"

To which Abdul, who always had the last word in our Indian affairs, heartily agreed. "A few rupees would be plenty, Memsahib."

Rupees it was, six shiny Indian dollars, that made Bulbul's eyes fairly pop as she raced home with them, almost forgetting, in her haste, the current haul of tinfoil.

When next we saw her, she was wearing a bright new necklace. It consisted of six silver rupees rubbed to a high polish, punched with holes and strung on a cord.

# 7
## Vagabond Fossil

"Old Bones," as we had come to call the specimen, was taking its resurrection very nicely. The freshly treated bone glistened in the sun, and the great ridged teeth stood out clear and sharp against the rock. The skull was in much better condition than expected. The lower jaw had turned out beautifully, and even the tusks, after repeated saturations of shellac, decided to hold together.

But—where was the body? Barnum made frantic efforts to find out, digging up what was left of the streambed in his search. He had Abdul digging, too, and even Tika, the mayor—poor groaning Tika who didn't know one end of a shovel from the other. Meanwhile I was pinch-hitting with Fasil in the cook-tent. Already, several cubic yards of landscape had been displaced without result. This meant more furious burrowing, with deep cuts radiating from the skull, in all directions, the excavations extending day by day.

"You'd think Solomon's treasure was buried here—the way you boys are sweating," I observed during one of their breathers.

"What's a little thing like Solomon's treasure," my husband countered, "unless it includes this critter's skeleton?"

As a matter of hard cold fact, there *wasn't* any skeleton—not so much as a piece of rib. The tired hunter finally had to admit it.

"But how do you account for that?" I wanted to know. Then, turning facetious, "Skulls usually come equipped with skeletons, don't they? Or were the prehistoric models different?"

Barnum fingered his chin thoughtfully. "Well, as I make it out, this fellow must have suffered a considerable loss of anatomy during his post-mortem wanderings."

42

"*Post-mortem* wanderings?" I echoed, thinking him surely joking. "Aren't ghosts slightly out of your line? I thought fossil-hunters were only concerned with the hard parts."

No comment. The man wasn't paying a scrap of attention. He sat gazing at the ground, his mind a million years away. He seemed to have drifted off into a trance.

Whenever this happens, there is only one thing to do—let dreaming scientists lie! It's well to have some knitting handy, or a book to read. I carried a small writing-kit for such emergencies and had answered some business correspondence and dashed off a condolence when the professor revived.

"What was I saying?" he inquired absently.

I put down my writing. "We were discussing skulls, sweetheart. How this one of ours knocked about after its decease, and how its body became lost in the shuffle. Remember?"

My scientist knit his brows, which makes him look sharp and scholarly. "Ah, yes. I was referring to the dislocation and obvious dismemberment of our elephant occurring between its first and second burials."

"You mean the animal was buried—twice?"

"Precisely. It's a typical case of secondary burial." To illustrate his point he squatted down and commenced tracing dohinkies in the sand. "This zigzag line represents our stream-bed," he explained. "And over there," his awl indicated the far end of the line, "somewhere upstream from our present position, the beast departed this life and became buried in the mud of the stream-bottom. So far so good. He might have remained there and in time petrified into a reasonably preserved *complete* fossil."

"I see."

Barnum heaved a sigh of regret. "But he didn't *stay* buried. Sometime after the flesh had decayed, strong currents washed away his resting-place and the force of the water bore the skeleton downstream. The various bones, no longer held together by tissue, went their respective ways. The skull became lodged and was reburied where we found it. God knows what happened to the rest of the creature; some of it undoubtedly was destroyed by water-action, other parts scattered helter-skelter along the stream-course and buried again." My husband thumped his thighs and rose. "That's the story, Pixie and—I'm stuck with it," he laughed. "Complete specimens are scarce as hens' teeth in this business."

"Anyhow, it's the bones you find that count; not the ones you miss," I noted cheerily.

"Right you are," came the reply. "Now, before we take this baby up, how about giving it a good old photo-finish? That's your job, you know."

I knew—although at the moment there was little to indicate what I was letting myself in for. Since school days photography had been a pleasant hobby with me. Once or twice, when I had snapped an especially good shot, I thought of it almost as a form of art. Now, it's just a pain in the neck. A rock, a ruin, a tree each has its possibilities, but a fossil in a photo doesn't look like anything at all, even like a fossil. Not only do most of them not photograph well, they just don't photograph.

It's one of those necessary evils. To quote the master— "A complete film record of the excavation is essential to our research later in the museum."

So, all innocence, I unstrapped my trusty Graflex, plus my trustier Brownie, and set to work, my husband massaging vigorously behind one ear. I thought I heard him snicker, but passed it off as imagination. This was going to be a cinch.

The initial step in such work is dolling up the bone; to wit, dousing it with shellac. This provides contrast and sets it off from the surrounding rock while still wet; the object then being to get in as many shots as possible before the stuff dries.

However, unless you are a confirmed scientist, you'll find it takes more than shellac to make a bone photogenic. There are several schools of thought on the subject. One maintains that the only way to get anything out of a picture of a bone is to put something into it—preferably something alive.

Being wide open to ideas, I tried rallying Taj to the cause. She was quarry-broken now and knew how to behave in the presence of rare antiquities. Abdul, too, was added for local color, and Bulbul,

Tika, plus friends plus camels for background. The resulting view showed error number one: too much setting; not enough bone. That's the way your experimenting goes.

Close-ups, you decide, are just the thing. Something with a touch of human interest showing the skull in all its glory, with Barnum looking nonchalant on the side; or romantic,—say, striking his best explorer pose, topi tilted back, pipe clenched in teeth, frowning darkly through his magnifying glass a la Sherlock Holmes.

A set of tools strategically placed around the specimen is never amiss. It reminds people that fossil-hunters have to work for a living like anybody else. A couple of whisk brooms, an awl, a pick, some chisels and mallets, a shellac pot or two and a messy plaster-pan should give a subtle undertone of activity to the scene. This type of print eventually ends up filed under, "Barnum and his tools."

If one hasn't succumbed to the dark-room blues at this stage, he returns to his original idea of snapping the fossil sans anything—except, perhaps, just enough rocky background for geological flavor.

That's what I did. I concentrated on the bone and blazed away from every conceivable angle—head-on portraits, close-ups of the dentures, trick shot along the tusk, some profiles, a precarious vertical from an overhanging branch. All of which, when developed, looked like nothing more than a study of a hole in the ground. There was a big gob of stuff in the center of most of the pics that we decided must be the specimen, but it wasn't very pretty. Petrified skulls rarely are. A few of those taken with the Brownie weren't too bad.

Once the detailed picture-story of our Spooky Hollow "find" was in the developer trays, Barnum made ready to take the specimen up. We cut some burlap sacking into long strips which, soaked in plaster-of-Paris, were wrapped snugly about our prize. This covered the exposed bone and, when dry, formed a hard protective casing.

The great head must have weighed close to a ton. It took all of us, including the camel men, to budge the thing. Once out of the hole, they lashed it to an improvised sand-skid and, with the help of the *dudh-wallah's* oxen, hauled it to a corner of camp. There, covered with a tarp, it stayed until boxed and shipped to the nearest railhead, along with other treasures Barnum was yet to find.

"A good beginning," was the modest way he described it. "Let's hope our luck holds out."

And why shouldn't it.

"You never can tell in this game till you go over the ground," he answered. "Even then it's ninety percent gamble. We might strike the 'Mother Lode' tomorrow, or go weeks without sighting another scrap of bone."

Since prospecting alone would tell the story, it was back to the work Barnum loved best—the hunt.

Each morning at sun-up, pick over one shoulder, canteen slung from the other, he started off on foot to comb the region in the vicinity of camp.

I would have gone with him and called it a make-shift honeymoon had he not been so set against it. "Prospecting's a one-man job," he said. "And a tough one. You let me bag the brute first, and then we can both go to work on him."

I had plenty to do in camp, anyway. That I told myself, was where a wife belonged—not galavanting all over the countryside looking for something that might not be there.

## 8

## Two-thirds of Siswan

But staying home wasn't easy. There were too many places to go—among them, the village of Siswan. Not infrequently, after I had seen Barnum off on his day's hunting and "done my duty" by Abdul, I'd grab a bite of lunch and head for town.

Siswan was really three villages in one. The largest consisted of Hindu merchants and weavers; another incorporated the scattered hovels of farmers; a third the hilltop eyrie of a group of Rajputs—each community isolated in its separate world of caste. These people might have lived at the opposite ends of the earth!

The town didn't offer much in the way of "sights." It was old and the years had treated it shabbily. Along the ancient main street lurched the wrecks of buildings rising from the heaped debris of ages, their interiors naked to the open sky, or, at best, protected by sagging beams and scraps of roof grown with grass and weeds. The crumbling wall, the broken column, the standing fragment of archway—all were there. And the sadness, too; for in spite of the dilapidation, there was a fineness of line, an art, which age and rubble couldn't hide. And as I walked the silent street there were ghosts, it seemed, hovering over the ruins.

What tales they could have told! Tales of old Siswan—of the life that was here in the fabulous days of the Great Moguls; tales of the market place and roadways teeming with traders and warriors, the princely trains of noble chieftains, the shuffling elephants richly-caparisoned with tall canopied *howdahs* rocking atop their backs, and the endless lines of dusty camel-caravans. Those were the days! Siswan held her head high then. She was a proud city known the length and breadth of the land, a great trading center and an important link in the long chain of forts that stretched across the Punjab. But all that was three hundred years ago. Now Siswan is a country village, content to sleep away the years that are left, and to dream of its youth.

I generally dropped in on our friend Tika Lal when visiting town. He lived with his family in one of the huge tumble-down houses, a somber place opening through heavy carved doors of Sisoo wood ornamented with hand- wrought iron tracery. It must have been a show-place once. Inside, the walls closing off the hall-like rooms were covered with frescoes, the brush-work of some master hand of long ago—painted peacocks, sacred bird of India, against a jungly background of trees and flowers, sportive tigers surrounding a rotund likeness of the elephant-headed Ganesh, patron of Siswan, while witches in hollow trees ogled a passing parade of jugglers, wizards, princes and bejeweled ladies, many dimmed by age and barely visible in the dingy light. Above, high wood-paneled ceilings inlaid with squares of glass!

There all evidence of former grandeur ended. Several of the apartments, where once some Oriental playboy sat in silken splendor, were given over to the storage of cotton. In others, amid deep drifts of fluff, sat low stools and ancient spinning wheels. Only the back rooms were lived in, and these seldom during the day. The family spent most of its time in the sunlit adobe-walled court at the rear of the house.

They had come to know me well at the Lals—the womenfolk, the children, and the old hound-dog that I was always amazed to find still in the land of the living. He was eleven years old, which meant that he had spent at least eight of them on borrowed time—a noteworthy feat in a town with as active a leopard population as Siswan's. I always remembered my own little Taj, and Abdul's remark, "Yes, Memsahib, dog is the leopard's favorite food."

Every detail of my first visit stands out clearly. Tika and I had just come from camp. When we entered the courtyard there was a sudden burst of childish fright as the tots spotted me and scattered, one screaming what might easily have been the Hindu version of, "Come quick, Mom. Look what daddy brought home."

Mom didn't come quick. I thought she would never come. Like the children, she had been completely unnerved by the unexpected arrival of a white woman in her home. My host had to clap his hands repeatedly before the members of his family gathered up sufficient courage to present themselves. They accomplished this by filing out of dark doorways one by one, as if going to the execution block.

In the lead came Grandma, head buried in her *sari*; then the younger women clinging desperately to their babies, the rest of the brood pretty well hidden behind their pantaloons. Last to join the hushed circle around me were some young girls and boys. A typical "One Man's Family" of India! All unwittingly, I had crashed the best *zenana* (women's quarters) in town.

Tika enjoyed three wives. As a Hindu he could have had seven. Wife number one was about thirty, the mother of three fine sons—and some daughters, the latter barely mentioned since girl children are none too popular in India. They cost too much! Because of the marriage "dot," every girl must buy her husband. And after two such weddings, plus dowries, the average father is well-nigh bankrupt.

Wife number two, the current favorite, was a sweet eighteen, with several male offspring already to her credit. Arrayed in pantaloons of dark blue homespun, embroidered vest and tunic, a short *sari* draped about her shoulders, she was a good example of what the well-dressed country gal will wear. The pearl-studded nose ring looked especially "chic," and her long black hair was braided with silver bells.

Half-portion number three was only twelve—too young to have anything but hopes. Her duties consisted mainly of helping the other two wives with the housework, at the same time receiving a liberal education in the art of making a husband happy.

The village or country wife in India has a far better time of it than her big-city sister. She has much more freedom. There isn't the strict seclusion to quarters, nor will friend husband be offended if she goes out without a veil. In many cases, she will even be found helping her man in his work. Then too, although neither so ornate nor luxurious, the country *zenanas* are more liveable, having a plentiful supply of air and sunshine. Some of them are delightful garden spots with the song of birds, the perfume of flowers and the graceful arch of trees over the walls.

Woman smiles in the country and small villages. She isn't merely a rubberstamp for childbirth, as motherhood is viewed in crowded centers where the husband's word is law and daily life is hedged 'round with multitudes of do's and don'ts, archaic customs, out-worn beliefs and medieval superstitions. Beyond the city gates these tyrannies stand for little, and woman assumes something of the dignity that is her birthright.

Nothing reflects this comparative freedom of rural women so clearly as that most popular of feminine institutions—the village stream. Where there is no stream, its place is taken by the community well. Whether stream or well, it serves as the social club, the sewing circle, the forum—focal point for all feminine interest, public and private.

Siswan was blessed with a wide, clear-running stream where each morning the fair sex gathered. You could see them winding through the trees to the water edge, their long white *saris* trailing to the ground. Ostensibly they came to perform the daily washing, though only the naive believed this their only reason. Normally, it was the busiest place in town, for there was always something, or preferably someone, to talk about. Of course, some laundering was done, if only for the looks of it.

The technique is quite interesting. After seating herself on a boulder in mid-stream, the Indian housewife grasps the unsuspecting garment in the left hand, gives it several vigorous dunkings, places it against the rock and whales the daylights out of it with a bat. To wring, the foot is placed on one end, and whatever remains following the twisting is slung over her head while she proceeds with the next. When the stack on the head impedes her arm action, she retires to the bank and carefully spreads the wash on the mud to dry.

Laundry finished, the lady proceeds to bathe herself, squatting down in the shallow water, knees jack-knifed under her chin, legs pressed tightly together. Off comes the *sari*, which she parks on a handy rock within easy reach in case of heart disease or fire. If husband is prosperous she uses soap; if not, plain water. She then washes the *sari* and dons it dripping wet. During the entire performance the lady hadn't shifted gears once—thereby demonstrating the limits to which Oriental womanhood will go to preserve her modesty.

Another point in favor of country living lies in the house-furnishings, of which rural homes have precious little. At the Lals, for example, all the secrets of housekeeping lay bare before you. Along the far wall stood earthen chatties containing the water supply. The cookstove in the outdoor kitchen was large and earthen. Scattered about it were several pots and plates of brass, this being the one metal that orthodox Hindus will eat from. In a corner were several stacks of what might easily be mistaken for *chupattis*. They formed the family fuel supply—dried cakes of cow dung. I was tempted to take a few home to Barnum labeled, "petrified pancakes." No Indian John L. Lewis has cornered this market as yet. To keep the home fires burning, all a woman has to do literally is "follow the herd."

House cleaning? Nothing to it. Leaning against a wall, along with the twin baby cots, the family bed is out for an airing—four legs on a wooden frame with nothing in between save some knotted rope. If you live in India long enough you become an expert "knot-dodger," painfully acquiring your own version of the rope trick.

Fortunately, the all-powerful sun is an effective germicide in the rural areas, thereby saving the price of a "bug charmer"—that most necessary character who walks along the streets of the larger cities tapping a big stick and crying his wares in a sing-song voice. His stock in trade is himself. For a few pice (farthings) he will sleep in your bed until the objectionable livestock is fed up. The secret of his profession, it is said, is an herb which the charmer eats in order to drug the bugs. India takes the cake for strange occupations!

During and just after the cotton harvest Tika's house was a beehive of industry, the silent rooms filling with the bustle and hum of a spinning mill in action. It was a cooperative business with the Lals, each member of the family a partner and everybody doing his share. Several times a day Tika and his boys brought home great hamper-loads of cotton from the fields which the womenfolk fluffed with long bows of buffalo hide after the seeds had been extracted. From them it passed on to more women and girls who spun it into thread on old-fashioned spinning wheels similar to our antique colonial models. Once spun, they wound the thread on bobbins that fit the village loom and took it to the dyer whose back-alley shop consisted of rows of large vats containing red, blue and green vegetable dyes. Then to the weaver went the dyed thread where it was woven into the coarse Khaddar cloth later sold in the bazaar.

Many a day, when business piled up and the courtyard hummed with the whirl of wheels, I became the umpteenth member of the Lal family; not as a wife, but as a baby-sitter, caring for the wee ones while their mothers worked. And I often thought, not how different, but how much like us these people were. Home folks are home folks the world over.

After my visit with the Lals, if time permitted I would return to camp by way of the farmer village, dropping off at the Gudyar's, the *dudh wallah's* place. I always enjoyed calling on them; they were much more fun than the Lals—perhaps because they didn't take life so seriously.

It was also the diplomatic thing to do. The citizens of Siswan were a sensitive lot, easily slighted, and there was considerable rivalry between the Lals and the Gudyars. As Bulbul kept my social activities under close observation, I was sure she knew every time I stepped into the Lal's house. To have visited the one and not the other would have been a faux pas indeed; someone's feelings would surely have been hurt, not to mention the possibility of our milk deliveries being cut off. To keep peace in Siswan and milk in camp, I didn't play favorites.

The Gudyars lived in a broad lush valley of pasture lands and wheat fields. Home was a cluster of grass-thatched mud huts, dirt-

floored and windowless but opening onto a brilliant world of sunlight and growing things. There was a certain "who cares" abandon about the place, as if they were too busy just living to keep
things tidy. One readily understood this on realizing that several
barnyard pets were listed as "family." These included a huge hump-
backed Brahman bull who looked like Satan, but had the disposi-
tion of a kitten; a water buffalo with twin calves, soft and curly,
with never a hint of the homely things they would grow into; an
off-white heifer of dubious ancestry, and sundry other bovines of
assorted natures. Chickens occupied the remaining space. Papa
Gudyar's herd of buffalo "milkers" he wisely kept in quarters not
quite so confining.

It was a wonderful thing, this living-together of man and beast.
Born together, they ate, slept, had fun and got sick together. The
animals grew up a little human, the humans a little animal—a good
arrangement all around, especially for the youngest boys who spent
much of their time swinging on the beef, wrestling a calf, or trying
to get a rise out of Mother Buffalo.

Here I learned where Bulbul took her singing lessons—from
the birds, no less. The neem trees surrounding the adobe compound
were alive with them; friendly creatures, swooping down to perch
on the backs of the cattle or mingle familiarly with the chickens.

The Gudyars possessed little by our reckoning. Yet, what they
had was fundamental—enough to eat, a fire for the cool evenings,
loved ones, plenty of hubble-bubble pipes to go around for both
men and women, clothes, four walls and a roof to keep out the
weather. True, they thrashed their wheat with a wooden stick and
ground their flour on a flat stone hand-mill. They were poor as
rupees went, but they had no use for money. All the needs of life
were there within the confines of their own farm. In a sense they
were rich, for they had that rare and priceless gift—the capacity to
be happy with the simple things of life.

"Ah, the poor farmers . . . they are low caste," their neighbors
on the heights would say, with great disdain. They didn't dream
that being low caste can be a blessing, and that the farmer, knowing he has nothing to lose, is free to make life what he chooses.

9

THE OTHER THIRD

Of the people on the far hill we saw nothing. They never came to camp; I didn't venture near their village.

When questioned about them, Abdul would simply say, "They are Rajputs," as if that explained everything. Further efforts to draw him out brought only mumbled somethings about Rajputs being very high caste, after which he closed up like a clam. Even Tika's ever-flowing vocabulary petered out to a trickle of ambiguous grunts whenever they were mentioned.

It was all very mysterious, and I had just about decided to let it go as such when, out of the blue one morning shortly after Barnum had gone for the day, I received an invitation from on high. Would I be pleased to pay them a visit?

Abdul immediately voiced his objection. He sounded quite upset. "Memsahib, it is unwise to go up there alone. Wait and go with the Sahib some other time. Rajputs are strange people. They have strange customs and ideas."

"Why Abdul," I laughed, "one would think you were afraid of them. Not to accept their invitation, now that I've visited the other two villages, would look definitely unfriendly. No, my mind is made up. I'm off to meet the Rajputs, come what may."

When he saw that I was really in earnest, he insisted on accompanying me—to my profound relief. It would be no small comfort having man-mountain Abdul by my side in case those *strange* Rajputs didn't happen to like me.

54

With my reluctant companion commending his soul to Allah, we began our winding climb to the top of Siswan's "Beacon Hill."

The moment we set foot inside the village a band of smelly Amazons swooped out from behind some adobe walls and surrounded us. Abdul let out a bellow and vanished into air. He reappeared, however, some moments later on the sidelines where he stood appraising me with anxious eyes. Not that he could see much, for at the moment I was completely engulfed in a chattering mass of female humanity, all trying to welcome me at once. It was the Rajput reception committee. Needless to say, I was "overwhelmed."

A few overtures of a getting-acquainted nature, and they escorted me to what appeared to be the village common, except that right in the center of it stood, of all things, a bed. To make matters worse the bed apparently was for me, and on it I found myself forthwith installed. Had I only known, I should have felt deeply honored. This enthronement on the community bed was one of the highest tributes they could show—a stranger—an honor usually reserved for visiting wizards only.

Instead, I couldn't help feeling a trifle ridiculous sitting there on exhibition, as it were, while the townspeople turned out to stare, eyes big as saucers. The men made a bold show of it, the children cried, the women edged in close to feel my clothes, skin, hair. All were looking, looking. That was just their way of showing interest, of course. Naturally, they were curious about a white woman. Chances were they never had seen one before.

Well, the feeling was mutual. I never had seen a Rajput before. To my amazement, I found that they were not at all hard to look at—in spite of their rags. Character and dignity shone through their poor exteriors, and there was something fine and alive about the eyes; something unwavering, piercing. The men, tall and spare, bore themselves like kings; the women with the easy grace of queens.

And why not? Were they not Rajputs—the people of "Royal Descent"? Did they not belong to that proud race that has given India her greatest monarchs, warriors and heroes?

None of this nobility was evident in their village, however. That more closely resembled their clothes, an untidy group of hovels looking more like piles of earth than homes. Nor was there any evidence of work such as one found in the other sections of Siswan. Rajputs neither sowed nor reaped nor spun.

They weren't lazy. It was their pride. Not for them the plodding menial tasks of field and village! They were born to higher, more stirring things; to the sword, not the plow. Should a Rajput, within whose heart burned the fire of a warrior, grub for a living in the furrows? He who was born to lead, a fighter lithe of limb, strong of arm, bold of spirit—should he sit in the market place plucking cotton? Since when had men taken to doing women's work?

They were men, these Rajputs. No mistaking that. Beneath the rags and dirt they were men—with the look of lions, and the hearts, too.

The climax of the afternoon came when one of my many hostesses offered me a drink of deathly-sweet goat's milk served in an earthen bowl and stirred with the finger. I tried to catch Abdul's eye over the rim hoping he could start a riot or anything to distract attention for a moment. But no! The big idiot just stood there, uncomprehending. I made a mental note to fire him as soon as we got back to camp, closed my eyes and, with as brave a flourish as I could muster, tossed it off! It was down! But—would it stay?

A split second later that detail didn't matter. Hardly had the bowl left my lips when it was snatched from my hands and dashed to the ground. I leaped to my feet. What had I done?

Then I remembered. This was another of those *strange* Rajput customs. Rajputs are very high caste, second only to the Brahmans. They deem their food and vessels sacred, instantly defiled by an alien touch. Because I was an Outcaste, the bowl containing the milk had become "unclean" as soon as I accepted it, and no amount of cleansing could remove the taint. So to safeguard themselves, they destroyed the bowl. There was nothing personal in such an action which was dictated entirely by religion. Should even the shadow of someone of lower caste, much less of no caste, fall across their food, it is considered contaminated and must be thrown away.

As if to rescue me from my embarrassment, a shaggy old man in flowing beard now made his appearance. Several younger men accompanied him, the crowd falling back at their approach. Grave of face, he paused before the bed, pondering me. Then he nodded to his companions and bade me follow.

I complied without hesitation. It was growing late and I didn't want to be the excuse for any more broken pottery.

We had gone a short distance down a crooked lane when I heard the thud of running feet behind me. Abdul drew up panting.

"What do you make of this?" I inquired of him. "Where is the old man taking us?"

"He is Chand, the village wise-man and teller of the people's tales. We have been asked to the story-tellings this evening, Memsahib."

In a small cleared space on the other side of an old wall the party halted and joined a ring of waiting natives around an open fire. Quickly and without ceremony the old man seated himself in their midst, motioning us to do the same.

For a long time the group sat in silence, all eyes on the aged man. The sun set. A wind stirred the dust on the rooftops. Night crept out of the East. Chand was in no hurry to speak. He, too, appeared to be waiting. With a long stick he poked at the fire, his mind lost in the flames. And then, as if drawing his theme from what he saw there, he began.

Slowly at first, and in a voice weary and old, the words came haltingly. But as the wood caught and the ruddy flames spread, his words came faster, smoother, till they fairly tumbled from his lips. Abdul leaned forward, translating.

His people listened motionless, as Chand spoke of their beginnings; how they were the fire-born race, the *Kshatriya*, the warriors of India. He told of the inherent democracy within the Rajput clan; how no one member was better than another, and that the humblest of them was kin of kings. From the earliest times, he said, wherever there was fighting and dying to be done, wherever the call to arms, there you found the Rajput. When that oft-repeated cry, "The Toorkh! The Toorkh!" went up through the land, who was there with banners flying and swords unsheathed to breast the mailed foe? The Rajputs—always the Rajputs! Ever have they stood, guardians of India's freedom, watchers at her gates; sometimes winning, sometimes losing, but fighting to the last like heroes.

He told also of men reckless in war, tender in love; of Prithvi Raj, Prince Rawal, and the courageous Rajah Sanga—names graven deep in every Rajput memory. For they were old tales first spoken by the famous Rajput bards, the troubadors and jongleurs of India, and they sent our minds spinning back through time to an age of chivalry and romance.

Before me rode the Rajputs of old—India's knights in shining armor—the Rolands, the Cids, the Charlemagnes, the King Arthurs and Richards of the East. And I wondered how many times the old chronicler had sat before his people like this, telling and re-telling their legends; and how many other old men before him had told and retold the same stories, keeping alive down through the ages the fierce pride of race in the Rajput soul.

It was dark now. The world had grown dim around us, the old walls of the village blotted out until there remained only the stars, the fire, and that ring of intent faces. The flames licked up from the burning wood, darting out, stabbing at the night as if to throw back the encroaching blackness.

In the flickering firelight a subtle change came over the listeners. The old men's eyes grew young again, it seemed, and the ragged backs grew straight. The faces of the young men glowed as they peered out across the night into the dear brave land of their dreams. They were yearning for the life of the warrior—the midnight raid, the headlong charge, the sudden ambush, the march and countermarch. They were longing to feel all the wild abandon of war that they somehow had missed—they who, on the battlefield, had no peer in all their world.

And the women too felt the magic of that night; saw their dreams take shape in visions, saw the mighty battlements before them, the armies locked, their men the bravest of the brave. Saw themselves. And they sat with quiet dignity—remembering. They, too, had made Rajput fame.

These were not a poor, broken people. These were warriors—proud, fierce, freedom-loving. Suddenly I realized why they lived as they did. It was because their lives were a waiting, their village an encampment, a bivouac, during a centuries-long interlude between wars. Tomorrow, some glorious tomorrow, they would leave this hilltop and move out boldly to meet the foe. For this they waited. For this they endured poverty and filth and sickness. Tomorrow they would march as Rajputs should.

Then the "Old One" was telling of Chitore, a name that fires the Rajput heart, a hallowed name for them like our own "Alamo," "Bunker Hill," "Bataan." For the glory of Chitore is the glory of the Rajputs—the proudest moment in their history. Chitore, the hill-fortress, the sacred stronghold where they fought and fought and died and died and lost and lost again to the might of the Mohammedan armies.

At mention of Chitore a quiver ran 'round the fireside circle; the fire seemed to blaze afresh. Amid a general tensing, pulses quickened.

Chand was a magician, his words a magic-lantern projecting the story of his people in panorama against the black night. I could almost see in that crumbling clay wail before me the bastions of the fortress, the Moslem hosts of Akbar the Great drawn up before them, the fighting—Rajput bravery and cunning against the overwhelming force of Islam; the daring escape of two thousand Rajput soldiers who, disguised as Akbar's guards, marched boldly through the enemy lines to freedom, carrying their own families bound hand and foot to look like prisoners.

With tenderness in his voice I heard him recount the tale of Padmani, Princess of Chitore and ideal of Rajput womanhood. She was the warrior-wife who engineered the great Indian Trojan-horse "double-play" and out-foxed the Mohammedan general Allah-ud-din.

It began with Padmani's beauty. Throughout the land its praises had long been sung and many heard and were content merely to listen. None dared hope ever to see that beauty—none but Allah-ud-din before whom all India trembled. Such utter loveliness was for him alone, and he could take what he chose.

Gathering his legions, he marched on Chitore, paused at the gates and delivered his ultimatum—one glimpse of Padmani, be it only her mirrored reflection—or immediate attack. This, Padmani's husband granted. But one glimpse of her and the passion to possess the Princess overcame the Moslem leader.

Moving swiftly, he took the husband prisoner and held him captive, under threat of death. The price for his life was—Padmani!

From the fort came seven hundred curtained litters—the perfumed palanquins of the Princess and her ladies. They passed by the sentries and on into the Moslem camp. Then at a signal, there came the rustle of heavy curtains, and out from the litters sprang, not the beauty of Chitore, but its strength; not maids, but men armed to the teeth and joined by the litter-bearers who had dropped their disguise and revealed their hidden weapons.

The stunned Mohammedans momentarily relaxed their vigilance. In that moment the captive Lord of Chitore was spirited back to the fort and into the arms of his beloved wife.

The speaker paused in order to let the words sink deep. We hung there, poised in silence, waiting. . . .

When Chand continued, his thoughts turned back to Chitore again—Chitore in its hours of despair, when the tide of battle had turned and the enemy came pouring through the breach to storm the citadel. He told of the lost hope and the desperate preparations for *Johar*—the mass-suttee—and of the great war-sacrifice of the Rajputs.

Johar! Johar! What terrible meaning lay behind that word! As the old man talked I saw the fort rising there, its center the funeral pyre, a vast pyramid of fire. I could see the women in their wedding garments huddled before it, the tear-stained faces, the fear-crazed eyes, the last look back at husband and life; then the mad leap forward, the screams smothered in flames. The children—too—the innocent lives just begun—snuffed out in that fiery bed with their mothers. And the cattle and horses, the tapestries on the walls, the princely robes, the art treasures, the wealth of centuries—all went into that blazing pyre. I could almost *smell* the holocaust, the burning flesh, the blood. I could almost hear the wailing of the living, the moaning of the dying, the clash of arms and crash of battlements. That was the way of the Rajput in defeat—the return to flame from which they, the fire-born race, had sprung.

They never surrendered. Always, when the fight was lost and the end near, there was Johar. As their world went up in smoke behind them, the remnants of the garrison grouped for a final gesture of defiance. They threw wide the gates and charged to certain death.

And when the victor entered, what had he won? Only blackened ruins, a smoking heap of rubbish, ashes—the sole spoils ever taken from Chitore.

As Chand brought his tale to a close, his voice grew thin and weak as it had been in the beginning. His face lost its animation and froze into the mask of age. He made a sign that he had finished, nodded to his audience, and departed. Slowly, in silence, the remaining Rajputs followed. At last only Abdul and I were left.

As we turned to go I noticed how low the fire had burned. There remained a few glowing embers.

All around was blackness. It was like the Rajput spirit, smoldering, feeding upon the strength of its past, waiting to burst into flame again when next India needed it.

I think back to that night in Siswan, whenever I hear of conflict in India, with the Rajputs riding again and the land ringing to the deeds of their daring. New tales will be told someday around new campfires—in the new India.

## 10
### "MAY YOUR HUSBAND OUTLIVE YOU!"

Drums. Drums throbbing in the distance, like faraway thunder; then louder, closer, as the canyon walls caught up the booming and sent it crashing down on our ears. With the booming came the blare of horns.

Could this be a holiday? Had the Rajputs taken to the warpath? Abdul would know.

Abdul, I discovered, had already climbed to the rim of the ravine above camp and was scanning the horizon through Barnum's binoculars.

"Ahoy up there! See anything?" I called.

He lowered the glasses, looked down, and yelled something. I tried to catch the words, but a sudden rush of sound behind me drowned them out. I whirled to find half the population of Siswan pouring pell-mell into camp, all in a state of jitters. Apparently they had heard the drumming. One of them was pressing through the crowd toward me, throwing his arms in the air to attract my attention. It was Tika. I waved back and started running to meet him.

Then, to one side, I spotted a topi bobbing along just above the brush tops. The helmet surged closer; the thicket shook, opened, and a disheveled and much-worried husband broke into the clearing in front of me.

"Hey!" he hailed, "what the devil's going on here? Heard the noise and thought it was you beating off another monkey attack. When the horns started in I didn't know what to think. Where's Abdul?"

"You know as much as I do," I answered. "Sounds like hells-a-poppin' somewhere. And the man who knows is at present up a cliff having the time of his life with your binocs." I jerked a thumb toward our servant who was in the act of taking off down the slope in the best Olympic style, his Turkish slippers with the retroussé toes serving as skis.

His eyes sparkled as he slid to a halt beside us. "Small procession with big noise heading this way. Pass through camp soon. The drums say it's a Hindu wedding party."

"A wedding!" the Browns chorused. Lights! Action! Camera!—and we both rushed on-location. It flashed over me that perhaps a vicarious honeymoon was better than none at all.

I had the up-canyon trail well covered with my Brownie when the first of the party appeared, a five-man band—two balloon-cheeked Gabriels and a threesome belaboring tom-toms. If what they were playing was a wedding march, it was a bad beginning for a *harmonious* married life. In the wake of the din jounced the nuptial carriage, a closed palanquin elaborately draped in red silk with long gold tassels, and carried on the shoulders of four brown huskies. Undoubtedly it contained the bride. In the open litter that followed reclined a handsome young man bedecked with flowers—the groom.

If only I had Kodachrome, I thought, centering the procession in my finder. I glanced over to see how Barnum was doing.

Not so well. He was setting up the Bell & Howell movie camera, the tripod, as usual, suffering from nervous collapse. Shots to the right of him, shots to the left, but the best shot of all was—Barnum himself. Clad in a pair of short shorts, he was making a valiant effort to unscramble the legs of the tripod hopelessly scrambled with his own.

It began as a slow waltz, swiftly swinging into an old-fashioned bunny-hug with Bell pressed close to his manly chest and Howell draped over his shoulder. They were doing nicely when, without warning, Bell did the split, came up with a high kick, and off went the explorer's topi. He stooped to pick it up, and then it was his shorts that did the split.

With things ganging up fore and aft, Barnum got mad and the sand began to fly. Man and machine were now slugging it out in the river wash. The champ made a left hook to the jaw; tripod crossed with a right. Champ was down! He was up. They clinched. Champ clamped a half. Nelson on his opponent and was about to take it apart when Taj dashed to the help of her master, bit the wrong leg, and all three went down in a heap.

By this time everyone had forgotten the wedding party, and I used up two rolls on this historic bout between Man and Thing. Just as I crouched to record the knock-out, Battling Brown came to, glowering up at me with— "What in Sam Hill are you doing? For God's sake, girl, stop the procession. I must get some pics before . . ."

But the procession already had stopped. Apparently the spectacle was too good for them to miss. The young groom had alighted from his dandi and stood viewing the mix-up with obvious concern, though, in true Indian fashion, without cracking a smile. From the bride's litter I heard a giggle and noticed the curtain move slightly. The villagers, too, had gathered 'round, Tika among them. I called to him.

"Tika, will you please welcome the newlyweds for us? Tell them we'd be honored to have them rest here awhile and accept the hospitality of camp."

The groom smiled as the message was delivered. Striding over to me, he bowed and in perfect English expressed his thanks. "But, tell me," he added, nodding toward Barnum, "what is the Sahib doing?"

I explained that my husband was merely trying to take some pictures of the bridal party, and inasmuch as he hadn't succeeded so far, would it be amiss if we snapped a few before they moved on?

The fellow drew back, eyeing the Kodak apprehensively. Finally, with a reckless shrug, he nodded approval, and went stiff as a poker. The sun picked out the gold strands in his pale green turban and brocaded vest over which hung garlands of jasmine twined about him by some pretty nautch girl at the wedding feast. An ornate belt circled his slim waist, and the bronze limbs were bare beneath the folds of his *dhoti*.

He wouldn't make a bad double for Sabu of the movies, I thought, as I maneuvered him against a suitable background and went to work. Around exposure six my subject limbered up a trifle, having discovered that the black box was quite harmless after all.

"How is this?" he'd ask, trying a new position. His favorite, and undoubtedly the one he hoped posterity would remember him by, was the "fierce" pose—chest out, arms akimbo, with a scowl that would have made Kali, terrible Goddess of Bloodshed, tremble in her boots. Next best was the ascetic look, the one he wore when he thought of himself as a *Sadhu*, a Holy Man.

We could have gone on indefinitely, but my films were limited. Furthermore, I had other subjects in mind—specifically, the little lady behind the bright red curtain. When asked if I might photograph her also, the groom beamed. As his gaze shifted to the palanquin, I caught the "lover" look—glowing, tender, sensuous. That was the look I'd remember him by.

Barnum, strangely silent till now, sprang to life. Out came his light-meter. "That's for me," he chirped. "I'll set up there opposite the dandi. As she steps out I'll snap her in front. You get her behind." He bounded toward the litter.

Suddenly, the groom's face went black. A loud defiant "*NO*" rent the air. "No, Sahib. No!" He confronted Barnum. "Sorry, but you cannot take picture of wife. Married Hindu women are *purdah*; no man except husband can look upon face. Only the Memsahib is permitted to take Picture." Then, in milder tone, "Sorry, Sahib. It is the custom."

Slapping another round into the camera, I scurried to the off-side of the palanquin. As I hesitated before the entrance, a tiny hand reached out drawing aside the curtain, and I thought there must be a little child inside. There was—a child hardly more than eight years old, her small frightened face peering up at me from among the cushions. Looking closer, I saw there were two of them, the other considerably older.

I glanced at the groom.

He smiled. "No, it is not as you are thinking," he said. "Only the older one is mine. They are sisters; it was a double wedding. The little one's husband will be along shortly. He had some difficulty keeping up with us." Through the curtain, the young man called to his wife in Hindustani, "Krana, come out. The American memsahib wishes to take your picture."

Slowly, with uncertain movements, slim little Krana stepped from the inner dark of the litter into the sunlight—a dazzling array of silver and gold. A pair of soft eyes and a winsome smile were all I could see of anything human.

She wore much metal. On her head was the last word in silver chapeaux—inspired, it seemed, by the top of a cocktail shaker. Silver bells wreathed her forehead. Her tiny ears bulged with silver flowers. Necklets and chains of gold and silver trembled sexily on her breast. Cuffs and anklets weighed down the small hands and feet. On every finger I noticed a ring, and on the thumb a large silver ornament that held—a mirror.

She was just conforming to the old Indian custom of wearing your wealth on your back. All the women do it. As there are no banks in rural India, the wealth of the family is kept on the woman in the form of adornment. When friend husband wants to put away for a rainy day he buys his wife a jewel, a bangle, or he strings some rupees around her neck. Krana sported enough solid cash to keep hubby in clover for years.

Of course the bride wore clothes, too: blue silk trousers, Jodhpur style, several sizes too large around the waist (with an eye to the future); white silk tunic; filmy pink head-veil; silver embroidered slippers. Between jewelry and clothes, one could hardly see the bride.

Just when I had the little lady ready to be snapped, two incongruous figures edged into the far background. They turned out to be a very old man on a very old horse; a nag so swaybacked the rider's feet almost scraped the ground. Bedraggled and thin, he looked like some Oriental Don Quixote just back from a tilt with a windmill. Looked as though he'd been weaned on a sour pickle, too.

Throwing us a dirty glance, he rode directly up to the young groom and commenced giving him hell. Photography temporarily forgotten, I hurried over to Barnum to watch the show. "Who is it, the boy's father?" I asked.

"Tika says it's the missing bridegroom—the eight-year-old's. He's bawling the other out for stopping here."

"Why should he object to that?"

"The old roué can't wait to get home. You can see the journey has been hard on him and he doesn't want to peter out along the way."

"Old Anxious" quickly worked himself into a fury. The only time his mouth and arms stopped was when he choked on his betel-chew. One last horrible exclamation and he whacked his horse and headed down-canyon. The young man spoke a few words to the litter-bearers, stepped into his dandi and called goodbye.

Hastily I pressed some flowers into the little bride's hand. "May you have many sons, and may your husband outlive you," I whispered. In India, these are the things a wife prays for above all.

The hand still clasped my bouquet as the company faded into the trail. Then I saw the old man ride up sharply. He bent low over the palanquin. There was a flutter of curtain . . . the flowers fell to the ground and the hand withdrew.

There, behind the curtain, the girl would spend the rest of her life in the veiled world of *purdah* women. After they had gone, we fell to talking.

"A rich wedding, the young couple's," Abdul commented.

"How much would you say?" I asked.

"Must have been a five-hundred-rupee dowry, at least. It was high caste. Notice the rich clothes? There was probably much feasting at the wedding. Fireworks, too. And such costly trappings on the palanquin! At a poor wedding, say only fifty rupees, the groom walks and the bride rides on the shoulders of some stout friend."

"Well, Krana and her sister will be glad when the trip is over and they can settle down in their own homes," I remarked optimistically.

Abdul gave a wry smile. "But it will not be in their own homes that they settle down. You see, when a Hindu girl marries her life becomes her husband's. The home she enters belongs to him, and the ruler is mother-in-law."

"Looked like the ruler was *husband* in the case of the little one and the old man. Maybe her troubles will be over in a few years—"

Abdul shook his head. "That's where you're mistaken, Memsahib. When he dies, her troubles begin. Life holds nothing for the Hindu widow. She must spend the rest of her days in mourning. She can never remarry."

Tika was quick to cut in, irritably, "The good Moslem gives a very one-sided picture of my people. Things are changing even now. Not everywhere do we treat our widows bad, and, as for childmarriage, it is an *abuse* of a very old custom. Long ago it was established to protect our girl children. Sometimes at birth, and even before, they were irrevocably betrothed, but never was the marriage meant to be consummated until they reached the age of maturity."

"It does seem a shame, though," I lamented, "that the sweet young things on the greatest day of their lives have to go around all covered up."

A queer light flickered in Tika's eyes. "Do you know who is responsible for the *purdah* in India?" He looked straight at Abdul. "The Moslems! Before they came our women were free. We had to adopt the veil in self-defense. The invading Mohammedans coveted our women even more than our lands."

Abdul chuckled. "True enough . . . but why do we argue this way?" He turned to Barnum and me with a helpless gesture. "That is the trouble with India today. The Hindus blame the Mohammedans, the Mohammedans the Hindus. When are we going to forget these old grievances that keep us apart? When are we going to think of ourselves as *Indians?*"

# 11

## KING COBRA IN THE COOK TENT

My husband is what is known in scientific circles as a "born collector," which is a nice way of saying that he's a highly specialized cleptomaniac. Although his talents lie definitely in the grave-robbing category, anything that comes under the general heading of Natural History is bound to arouse his pack-rat instinct, as he calls it. Whether alive or dead, if it isn't nailed down or marked "private" it's a potential museum-piece for Barnum and, therefore, legitimate loot.

The psychological value of this omnivorous attitude has its points; Barnum never comes home empty-handed, never feels discouraged. When business is bad in the bone line, he'll collect anything handy.

"Best fish are caught on a deer hunt," is his working slogan.

Among the many things he collected at this time was a whopping big lizard. He sauntered into camp with the thing trussed on his shoulder. It was two feet long—alive and wriggling.

"And what are we to do with this?" I inquired. "A playmate for Taj, I suppose?"

"Pickle him," was the grim answer.

But in what? We rummaged vainly through our equipment for a jar large enough to hold the animal. Honorable Liz finally ended up, properly pickled, in the "commode."

Snakes, too, found a place in the old collecting bag. One resulted in Barnum having a mouse named after him. It was a two-in-one haul. The snake lay stretched full-length across the trail, a suspicious bulge in its middle, when Barnum came upon it. Curious, he opened his jack-knife and performed a Caesarean on the creature, removing the lump. Behold, a mouse—a very extraordinary mouse, as it turned out. Shortly after he had dispatched it to the Museum, back came a wire of congratulations. The mouse was a new species!

We broke even with another snake, gaining a magnificent scientific specimen, but losing a pet. The pet was Mimi, a little hen Abdul had brought from town and intended to fatten up for a Sunday feast. The fatter it got, the cuter. As a result, our chicken dinner was postponed time and again. Mimi would undoubtedly have gone right on laying eggs for the duration—had not a hungry king cobra visited camp one 2 A. M.

Loud cackling in the cook-tent awakened us. Then came the patter of feet outside, and Abdul's voice: "Hurry, Sahib! Big snake!"

Barnum and I were up in a flash, and off—a weird procession in the dead of night; one torch-bearer, one sahib in striped pajamas armed with flashlight and pick, one scared memsahib in bedclothes. It was sneak—sneak—sneak. Shushes. Then out with the torch, on with the flash. . . . A huge black cobra lay on the cook-tent floor, still as death. The flashlight beam ran along the fat curving body to the bulging head, and there, protruding from the open jaws, we saw the legs and rear-end of Mimi.

Barnum lunged forward, bringing the flat side of his pick handle down hard just back of the ugly head and holding it there, his entire weight on it. Instantly, the chicken was disgorged. The snake, head pinned to the floor, writhed and lashed out against the tent sides like a great whip, scattering pots, pans, bottles and jars. Barnum pressed harder, beads of perspiration breaking out on his forehead. We waited, breathless, as the death struggle went on. Gradually the thrashings grew weaker. After what seemed like an hour—possibly five minutes—the cobra lay still.

Quickly, Barnum reached under, closing his fingers around the broken neck. "Alcohol!" he panted. I slid a large enamel pitcher over to him. Still quivering, the snake was held high and with a "swoosh" dropped into the alcohol. A last wild convulsion as the great muscles tightened, then relaxed . . . and "the sahib with the charmed life" became the talk of the Siwaliks.

In the days to come he added to our collections minerals, mammals, bugs and birds and innumerable samples of plant life. In short order we had enough of the flora and fauna of the region to start a museum of our own. But nary a bone.

Never had he seen such a sterile waste in all his years of collecting, Barnum admitted. He was beginning to wonder if our elephant skull wasn't just a flash in the pan.

"Got to change tactics," he announced. "Spread my prospecting over a wider area."

That meant the hills—long treks, working away from camp three and four days at a stretch. Again, I hinted that I was available as a travelling companion. Again, he reminded me that prospecting was a one-man job.

Jenny, the roan, was limbered up. Barnum looked over his maps. Then the two made tracks for the badlands, a bedroll and several days' rations strapped to the saddle.

"Good luck, darling," I called after him. "Here's hoping!"

I didn't have to hope long. He was back that same afternoon, charging into camp like a ball of fire. Leaping from his horse, he grabbed my arms and gave me a resounding kiss.

"Lordy!" I chattered. "You must have made a strike to give out with a kiss like that."

"I have!" he cried. "I have! Came on it about two hours ago—the most beautiful leg you ever saw, sticking out of a hillside and labeled, 'come and get me.' If it doesn't lead to one of the best 'finds' of this or any expedition—well, Pixie, then I'm no prophet. Lying there pretty as you please in good shape, too, from what I could see of it." Saying which, he rushed over to the provision tent and did a sort of swan dive into the tool kit.

"What kind of a beast is it?" I ventured, standing clear of the entrance.

"Don't know yet," came from the depths. "But it's as big as a house. Got a limb bone like a rhino's—only it belongs to a reptile."

"Maybe it's a dinosaur," I piped hopefully.

My husband emerged from the tent looking like the lone survivor of a rummage sale. "Dinosaur?" He laughed. "The dinos' were all dead as dodos long before these Siwalik rocks were formed."

"What then?" I continued, hoping to draw him out.

No success. Scientists are like that, cagey and noncommittal until they have the facts—the hard rock facts. My scientist wouldn't even wager a guess yet. Hadn't seen enough of the bone. "But," he added, "I've a feeling we're in for a surprise. Some pretty astounding things have come out of the Siwaliks."

"When do I get to see this beastie?" I asked.

"Right now. Saddle up that nag of yours and we'll get going. 'Bout time my wife learned something of her husband's work."

"Lead on, MacBones," I yelled, giving Taj a last-minute hug.

A half-hour later we were riding down the canyon in a cloud of dust and "wah-hoos." Fording the stream, we climbed out of the ravine and headed cross-country.

Barnum pointed beyond the sparse upland to a distant line of hills. "There it is," he said, "on the far slope of that ridge, just below where you see the notch."

Our mounts settled into an easy lope, and, almost before I knew it, rounded a rocky shoulder. We were there.

# 12

## BEAUTIFUL LEG

The beautiful leg turned out to be a lower left fore-limb bone. It was nineteen inches long, stubby, thick and not very shapely, but my husband couldn't have been more excited had it been one of Betty Grable's.

"I'm betting there's a peach of a specimen on the other end of that," he gabbled, "and we're going after it."

"We?" I bounced back, incredulously.

"No one else, partner. You and I are going into seclusion for a while." He motioned to several other places in the bank where small patches of bone showed. "There's enough material on this site to hold us for some time. We'll set up shop right here on the spot; nothing elaborate, just an overnight shelter. Abdul and Taj can take care of home base. You with me, Pixie?"

"And how!" I responded. But I wasn't thinking the same thoughts he was. Mine leaned more to the honeymoon idea. How heavenly it would be to have Barnum all to myself for a change.

Ah, the hope, the stillborn hope that wells within a young bride's breast; hope doomed to frustration in my case by a conniving fossil! Of course, I should have known by now that the beast in the bank had prior claim to my husband; that a mere wife must play second fiddle—if she fiddles at all. I had a leg, too. But, unfortunately, it wasn't petrified. And so, because it takes at least a million years of burial in river, lake, or ocean sediments, and sometimes volcanic ash, for a normal, healthy leg to petrify, I forgot myself and pitched in for dear old American Museum.

Now, as any honest ghoul will tell you, digging up the dead is no cinch. Once a promising cadaver is located the question immediately arises, how to break the seals on its tomb without breaking the contents. Fossils, for all their antiquity, are extremely fragile; to excavate them requires a light touch and an understanding heart.

That was why, those early days at the dig, I was kept strictly in the spectator class, permitted only to watch while my husband showed off with pick and shovel. This, he called "removing the overburden"—the layers of rock overlying the bones.

Then he returned to the leg and commenced following it into the bank, feeling his way slowly by awl-point, chipping away the clay with mallet and chisel.

"Observe closely," he kept reminding me. But all I could observe was the seat of his pants and two elbows going like mad. Sometimes not even that for the dust. Head and neck were lost to sight.

This is the traditional pose of all self-respecting bone-diggers, and, although it bears a superficial resemblance to the stance of a football center in the act of snapping the ball, more than likely it originated with the ostrich. Of course, fossils have been—still are— excavated from the prone, sitting, kneeling and squatting position, but the "ostrich" is by far the most effective. In this posture a fossil-hunter is dead to the world. Come earthquake, fire, flood or wife . . . on he digs, oblivious. And when night finally falls, he looks up startled and thinks a storm is brewing.

When next I had a peek at our specimen, Barnum had laid bare the leg clear up to the pectoral girdle—shoulder blade, to you.

"What happened to the foot?"

He granted me a hasty glance. "That was eroded out, my dear. Erosion is the curse and likewise the boon of the bone-hound. A little of it uncovers just enough for the specimen to be detected; too much, and the treasure is gone forever. That is why we seldom find a complete skeleton."

On the seventh day we rested, lolling under the awning we had rigged up over the dig, and talking.

"I'm proud of you, Pixie," he announced, out of a clear sky.

"Whatever for?"—this when I'd recovered my tongue. All I had done the past week was stand around and look. And I told him so.

"That's why I'm proud of you," he said. "Most women couldn't wait to dive in there and hook that critter out. The most important thing in this business is patience, and you have a lot of it—for a woman," he added, archly. "I'll make a bone-digger out of you yet."

"When?"

"Tomorrow." He smiled indulgently. "Tomorrow I'm going to let you cut up—some burlap sacking."

"That will be just ducky," I retorted, unable to contain my elation. "Do you prefer your burlap cut straight, on the bias, or with scalloped edges?"

My husband smiled, then grew serious. "We'll be needing plenty of stripping, comes time for plastering later," he said.

Later was a long way off. Meanwhile, I had graduated from gunny-cutting to tool-holding. Naturally, I hadn't yet reached the state of grace necessary for the actual handling of bone. But Barnum compromised a little and we invented a system that enabled us to work together, anyhow. Like a well-oiled machine our movements harmonized, he with pick, awl and whiskbroom, uncovering, I with shellac pot and brush, working in the clear liquid preservative till the bones shone like satin.

Side by side we labored, slowly, carefully opening the ancient grave. And here beneath the burning suns of India the working team of Brown & Wife was cooked up—to continue through the years.

We were making wonderful headway when a shout from the master brought me to my feet. Peering over his shoulder, I saw a rounded border of bone which, as it was further exposed, took the shape of the flared rim of a gigantic army helmet.

Barnum ran his fingers thoughtfully along the cleared edge, pondering. "Know what I think?" he said at last, his face brightening. "This is the under margin of a huge tortoise shell. Looks like we've bagged a *Colossochelys*—a *Colossochelys atlas*, sure as shootin'."

"Yes?" I said, gingerly, not knowing whether to cheer or be solemn.

His explosive, "Good Lord, woman, where's your enthusiasm!" showed me that I should have cheered. "This baby's about the biggest thing in land turtles you can find anywhere," he went on. "That's what the name *Colossochelys* means—a colossal chelonian—a tortoise—the grand-daddy of them all."

"What does the *atlas* stand for?"

"An old legend is probably responsible for that. The ancient Hindus believed the world was upheld by a giant turtle, a sort of reptilian Atlas, and the myth undoubtedly figured in the naming of the first-discovered specimen of this type."

The turtle that supported the world! Sounded like an American species to me. "Lucky the SPCA wasn't around then," I muttered. "How long had this been going on?"

"Right from the beginning—the Hindu genesis, I suppose."

"Surely our specimen here isn't that old!" I exclaimed. Barnum laughed. "No fossil goes back that far—to the beginning. This turtle lived only about a million years ago—during the Pleistocene Period of earth history, that is."

"Some age!" I commented.

"Not as turtles go," he told me. "They're one of the oldest families in the animal kingdom. Pedigree goes back almost two hundred million years to the Triassic Period . . . that was even before the time of the dinosaurs, broadly speaking." My husband sat back, contemplating his pipe. "Amazing creatures—turtles," he went on. "They have survived longer than practically any other backboned animal, unchanged very little in design since Nature invented them. The original model evidently couldn't be improved upon much."

More digging along one side of the shell, the clay wall receding before the quick sure strokes of the pick. Deeper, wider grew the quarry. Close work, now, with brush and awl. Cutting the rock cover away further revealed still more treasure. A pocket of small bone nubbins—the right forefoot. A stubby paired shaft, a joint, a massive curving of bone—the right foreleg. A gap—empty clay. Time out for a stretch and a breath of breeze on the crest. Back to work.

Bone again!—slender, fragile, spool-shaped forms linked together end to end. "Neck vertebrae," Barnum observed, "the cervicals."

We had just laid bare the third of these when the worst possible calamity befell us, barring actual loss of the specimen. We ran out of shellac. What we had thought the reserve supply turned out to be—tomato juice.

An awful silence fell between us; a silence which seemed to imply that it was up to me, somehow, to save the day. Somewhere in the world there must be more of that substance so precious to bone diggers. But where?

Grabbing a bridle, I mounted Jenny and tore hell-bent back to camp. No shellac! Said Abdul, "You took the last with you, Memsahib."

Into Siswan I headed, my thoughts flying back to the third cervical lying out there exposed to wind and weather without benefit of shellac. Time was growing short. I could see Barnum tearing his hair and shouting, "Shellac, shellac. For God's sake, shellac!" I dug my heels into the roan's belly. We flew, her hoofs pounding a tattoo on the sun-baked earth. Shel-lac—shel-lac—shel-lac, they said. Down the main street, they thundered—to Tika's.

"Shellac," I panted. Our faithful friend took up the cry as we flew to the shops. To one, to another, up the street, down the street, in and out of the lanes we sounded our refrain. And the dreadful echo kept roaring back, "No shellac. No shellac."

"What now?" I asked, close to panic.

Said the merchants, in unison, "Fish glue!"

I bought them out, four pots in all, and galloped back to Barnum. The third cervical was saved!

Our big job was the shell, the "carapace," as the scientist called it. Uncovering that required no abilities other than a three-way stretch and a boarding-house reach. But at the end we felt as if we really had accomplished something. For, with the monstrous shattered dome rising from the center of the pit ringed by the smaller bones of the skeleton and by disconnected scraps, the whole of the specimen lay exposed. Here before us was seven feet of prehistoric tortoise which had weighed 2,500 pounds in life—the largest land turtle then known to Science.

In time the dead was ready to be plastered. So up the slope we hauled the large open pan, water, plaster-of-Paris, gunny. Included also were rubber gloves, safety goggles, old clothes and hats that came down over the ears. From which you may gather that plastering is a very messy business. Once you've mixed the stuff, you have to work fast before it sets. As a result, when finished, you're pretty well plastered yourself.

But Brown & Wife heaved to with a will and let the splatter splat where it had a mind to. I did the mixing, dunked the stripping, gunked them up and fed them one by one to the chief, à la assembly line. Over the tissue-thin sheets of rice paper used as separator next the naked bone, he slapped the soaked burlap, molding it to the bone, kneading it gently into the hollows and cracks with his fingers. On limb bones and larger sections of shell he reinforced the bandages by splints made of branches.

Plaster hard, the skeleton was now sectioned off into nearly equal blocks, each turned bottom-side up. Some chisel work followed. Very little, though, as sufficient rock had to be left on the undersides of the bones to cradle them in shipment. A few more strips of gunny, a brace or two, and our *Colossochelys atlas* was completely encased, ready to travel first class to America.

"That does it," said the boss.

We both collapsed. When we came to I looked at Barnum. Barnum looked at me. We both looked at the specimen in its dead-white wrappings.

"I feel great," he exclaimed. "How 'bout you?"

"Never better, Mr. Bones," I replied with what I hoped might pass for unrestrained enthusiasm. But had he chosen to add, "To heck with fossils," he would have taken the words right out of my mouth.

No leaping, squealing Taj welcomed us back to camp on that triumphant return. Taj was dead—killed by a leopard some nights before.

# 13

## Monkeyshines in Simla

Bone-hunters are a clannish lot. Let two of them come within five hundred miles of each other in the field, and they are drawn together as inevitably as turtledoves in spring.

When Barnum learned that India's crack geologist, Guy Pilgrim, was working the country to the north of us the entire future of our own expedition suddenly depended on meeting him. It didn't matter that the man's exact whereabouts was unknown, or that reaching him would require a long and arduous journey.

"Any road is short when there's a fellow-explorer at the other end of it," my husband remarked. "Especially when Pilgrim is that fellow. He knows the rocks of north India better than any man alive. A half-hour's talk with him would be of inestimable value."

"And how will you go about locating him?"

"The same way I'd track down a fossil," Barnum explained, a scientific gleam in his eye. "From last reports he was somewhere around Bilaspur. I'll head there, then make for the nearest bone-bearing rock formation and follow it along till I find him. I figure a week to ten days should do it."

Fleeting visions of a possible honeymoon at last—misty and moonlit—rose before me. "There wouldn't be room for one small wife?" I angled. "Just five feet two?"

He regarded me from under arched brows. "Maybe. But only as far as Simla," he hastened to add. "That's the end of the rail-road. From there on it'll be rough mountain-trail sixty-some miles into the Himalayas. You could stand a few days in a comfortable

79

Simla hotel. The rest will do you good. Abdul can take charge here while we're gone."

Visions of a possible honeymoon—misty and moonlit—vanished.

And just as well, perhaps, for Simla, summer capital of India, was hardly the place to bill and coo in winter. It had the deserted, forlorn look of Coney Island in January. With the exception of a few natives and merchants, the population—mostly British officials and government workers—had long since departed, going south to New Delhi.

My hotel was quite in character—a huge barn of a place perched on a ragged brown mountain and featuring rickety staircases, squeaky floors, draughts, and long dark hallways. Behind the desk sat a myopic character in horn-rimmed spectacles who looked like a gargantuan frog, with a deep batrachian voice to match. He wasn't native and he wasn't English, nor German nor Polynesian. He could have hailed from Mars. Somewhere in his past East and West had evidently met and mixed, both losing all identity in the mixture.

As Barnum and I entered the inn he leaped out from his cubby hole and commenced bobbing around like a jack-in-the-box, bowing and scraping and grinning from ear to ear.

"You wish room—you wish?" he kept repeating. It was more of a plea than a question. We couldn't say no.

"So happy—so happy," he chortled, scrambling back behind the desk and opening the dusty register.

Having installed me in a comfortable room with fire, and made certain that my every wish would be gratified during his absence, Barnum made a bee-line to the livery stables and was off to the region of snows with woolies on and two heavy blankets in his saddle roll.

"Chin up," he said, kissing me goodbye. "I'll be missing you."

I watched until he turned out of sight into a grove of pine, then suddenly I knew what it must feel like to be widowed.

The hotel clerk's name, I discovered, was Aloysius. He'd been raised as an orphan in a Christian Mission—parents unknown—and for the past several years had been employed as caretaker of this hotel during the off-season. In summer he served as handy-man.

"Summer is the time to visit Simla, Madame. Not now," he remarked. "Now it sleeps like the bears in the caves up the mountain—and who is pretty when they sleep?" His face executed a weird grimace. "In the hot months, when the rest of India bakes and steams, it is cool and sweet here in the hills.

"Everyone who can comes to Simla then. In March it is they start to arrive, loaded with trunks and furniture and children. It is hard even to get a seat on the train, and many come by automobile and truck. Nowhere is there so much life, Madame. There is tennis playing and horseback riding and swimming in private pools. Men go on hunting parties, or fish the streams. It is good here then." His eyes bugged. He was like a starving man talking about food.

"Did you know we have a racetrack?" he continued. "And a cricket field? Oh, yes, much sport. You see, the Viceroy lives here in the summer. He has a lodge up on Observatory Hill. And many officials and army men are coming and going all the time. There are people—people—crowds—all kinds."

The man paused and gave me a searching look. "I like people, don't you?" he asked, with an earnestness almost pathetic.

I nodded.

He went on. "That is why it is so hard to live here in the winter. No people! The natives seldom come up from their part of town and there is nobody to talk to except oneself—or the monkeys."

"Monkeys!" I echoed.

"Oh, yes. When they are here, which is only in the daytime, they occupy the west wing where I have my room. Great company, Madame. They will call on you, too, have no doubt, as soon as they discover No. 14 is engaged."

The great discovery took place in the frosty dawn of the second morning. I was awakened by something picking at my window screen. Glancing out from under the covers, I beheld a shadowy form on the outside sill framed against the early gray light. My first thought was that it was a jackal, a pack of whom had held a howling good conclave below in the ravine a few hours before. But the unmistakable hunch of the shoulders and the long hairy arms spelled simian. The beast was trying to work the screen loose.

"Hey, you. Beat it! Go on home!" I called out in the fiercest possible tones, and dove back under the covers. When I looked again, he was gone.

But not for long. I was teetering on the edge of another blissful doze when a new sound snapped me awake. This noise came from the door, and I turned over just in time to see one of the afore-mentioned long hairy arms reaching through the open transom. The bed clothes flew as I sprang to my feet—too late! Already Mr. Monk was on the inside of the transom, leering down at me. A moment's pause and he leaped lightly to the floor.

Frozen, we stood eyeing each other. He wasn't a bad-looking brute. He rather reminded me of our Siswan breed, only he was somewhat larger with a mangy reddish-brown coat badly in need of a touch of henna. It was then that the clerk's racial background struck me forcefully. Somewhere down the line he had a monkey in his woodpile. Though, on second thought, haven't we all? At least so Science says. It isn't hard to believe in India.

Such thoughts brought small consolation in my present predicament, however. The ape appeared to be making faces at me now. He took a step nearer. Desperately, I looked about for an avenue of escape. The monk blocked my way to the door and only a Sabine would have used the window to the rocks below.

Something I remembered as a child occurred to me then. If you are ever cornered by a wild animal, the something said, show no fear. Look the creature straight in the eye and stare it down! I did. But so did the monkey. Apparently, he didn't think much of the idea, for, after taking time out for a good scratch, he shambled right on past me to the dresser, which had been on his mind all along. He calmly, methodically picked over the various articles on top, then went through the drawers.

While their contents engaged his full attention, I snatched a blanket and slipped out. Believe me, the management would hear of this! At the end of the hall I glanced over my shoulder in time to see the varmint emerging from the room with my hand mirror.

This was too much! Emitting a war whoop—genuinely blood-curdling, as I had suddenly lost all fear—I took after him. He saw me coming, put on speed and skittered around a corner. I negotiated the turn safely, but when I looked he had hopped into a laundry chute and disappeared. Returning to my room in vast disgust, I dressed and descended on the desk clerk like a lady—with murder in her heart.

My harrowing experience registered exactly nothing on Aloysius. He simply shrugged helplessly and said, "It is like this all winter—every winter. Generally in the hot months the monkeys keep to the slopes of Mt. Jakko a few miles away where they live in an abandoned temple with an old Hindu Holy Man who looks after them. There is an entire colony of them. As soon as the people leave in the fall, though, they start coming to town. Food is scarce in the hills in cold weather, and they know there are always warm rooms and plenty of scraps in the hotels."

"You mean you deliberately keep open house for these creatures all winter?" I asked in astonishment.

"It is the custom, Madame. They do no harm; there is no cause to fear them. They average only about three feet in height. The older reach perhaps three and a half. Once in a while they may take a dislike to a certain person and pelt him with stones, but that is rare. They like most folks. The natives have a deep regard for them, too. Especially the Hindus who, as you probably know, have a monkey god named Hanuman. There is a special temple to him down in Benares."

Beginning to feel that the human species was a step *down* on the evolutionary ladder from the anthropoids, I went back to my room. Guests awaited me. My erstwhile Romeo of the dawning had returned with his family— Mama and two kids. The place looked as if the American Legion had held a convention in it. The family had been into everything, and now were simply raising general hell.

In utter resignation, I sank into a chair and watched. Fortunately, I had brought few clothes on the trip, for those few looked as though they had just been through Macy's basement. How Father had managed to get the elastic out of my undies I'll never know.

Yet there it was, with one of his brats chewing on one end while little brother tugged at the other.

Father was currently doing stunts—swinging from the chandelier and bouncing on the bed. Mother, a quiet little soul, sat on the floor slowly and deliberately picking over herself.

I finally managed to lure them into the hail with some apples, locking myself in to await the room-servant and tidy-up.

On the floor where Mama had been sitting was an empty tin of sugar-coated Cascaras. Where the pills had gone I learned the following morning. As I kept the transom closed nights now, Father Monk awaited me out in the hall. He wore a hurt how-could-you-do-it expression on his face. Wife and kiddies were nowhere to be seen. The Cascaras!

In time, my presence in Room 14 became common knowledge among all Simla simians, and I suppose you could count on your hand those I didn't play host to during my stay. For the most part, though, they gave my quarters a wide berth. That was considered Father Monk's territory. Friends of the family and a grouchy old buck who appeared to boss the whole colony were the only ones he permitted around.

Each day toward evening, before they left for their mountain top, I placed two large plates of dhal on the porch for them. Usually the old fellow would try to monopolize both plates, holding down one while he ate from the other, and cuffing the little ones if they got there first. The dhal provided a welcome addition to their diet. Their mainstays of fruit, seeds and insects were plentiful in summer, but at this period a monkey depended largely on handouts or an occasional raid on a pantry.

It was surprisingly easy to grow attached to them. Often, in the hall, as I went to and from my room, a clammy little hand would be slipped quite unexpectedly into mine and we'd walk along together. Sometimes, as I sat on the wide verandah reading, Father Monk would perch on my chair arm and play with the pages, jabbering about God-knows-what to me the while, and knowing I couldn't talk back. This was okay with me until one day he started picking over my hair, the way he did his own. Then they all got out. They didn't even get bread and water that night.

They made up for it, however, the day I received word that Barnum was on his way back. I celebrated by throwing a party for them. The refreshments consisted of bread soaked in beer. They went wild about it and I had all I could do to keep them supplied. It took very little for them to feel it, at which they became almost human in their antics—staggering around in a tipsy way, throwing their arms about and slapping one another on the back.

The only one who did not react happily was Father. He stole off by himself to the bureau top. There he sat with his back to the crowd, trying to drum up acquaintance with his reflection in the mirror. Reaching out, he would attempt to feel the image, then draw his hand slowly away and peruse it reflectively.

Anxious to learn what Father found so absorbing, Mama Monk swung up beside him and gave one look in the mirror. What she saw was *her* own reflection, but thinking it another female with whom her man had been flirting, she screeched in jealous rage and tore at the glass with hands and feet. That pretty well broke up the celebration, Father beating it back to Mt. Jakko in disgrace.

# 14
## "WATER BABIES"

Of course there was a more human side to life in Simla. Aloysius had been trying for days to get me to see the town. Eventually he succeeded.

A rickshaw is the best way of getting around. The city being built on a vertical plane, its narrow, crooked streets twist, dive, climb, again and again double back on themselves in following the sharp contours of the "ridge," as the mountain is called on which Simla hangs.

Simla is British, not Indian. It has been British since the end of the Gurkha War in 1816. In all probability it will retain its British character for a long time, regardless of the flux of politics and the ascendency of native power. The names of the cottages and villas carry the flavor of the place— "Battsley," "Waverly," "Snowdon," "Rothney Castle," "Squire Hall," "The Manor." I half expected a blighty bobby to pop out of some bloomin' alley as we went barging along. Then there was "Glengarry," and "Erin Villa," and my heart skipped a beat as we passed a marker reading "Brooklyn."

But, as all were shuttered and locked, and their owners in far distant New Delhi, we directed our runners to the native sections where, as Aloysius said, there were some real live people.

The natives here are different from those in Siswan or any other part of India. They are mountain people—hardy, happy folk with a vitality unknown on the hot plains or in the jungle. Some are small and squat with long black hair and round jolly faces. Some have a Mongolian cast to their features. People say they are Tibetan. Still others are tall and wild-looking, their thick black locks bobbed short on their necks, with gold earrings hung in their ears.

Many have come to Simla from far back in the Himalayas, along the old road that leads out of inner Asia into Hindustan. Migrating humans have followed it since before recorded history, just as today it guides those from hard cold regions to the softer climes of the south.

The tasks these men perform are almost beyond belief. One sees them trudging along trails, bent double beneath loads that would crush a horse. A heavy beam fourteen feet long and a foot thick carried on a man's back is an ordinary sight, and the chances are he has borne it for days over mountain and through snow and rushing stream. His possessions will be tied to one end—a coarse blanket to keep out the frost, a brass cooking-pot, a small skin filled with water, a sack of rice.

Practically all freighting is done on the human back. Shortly after the train arrives, you can see porters struggling up from the station, one of them, perhaps, juggling a trunk, valises, baskets, hat-boxes and what-not. Others are plodding out of town into the hills shouldering great sacks of grain or coal, or tanks of water. Six men will strap an iron girder to their backs and haul it miles through mountains, walking sideways where the trail is narrow.

The lives of their womenfolk are hard, too. Once when Aloysius accompanied me in his rickshaw we happened on a group by the roadside using small mallets to break up rocks for cement. As each filled her bucket with the broken stone, she would carry it to the cement mill at the top of the hill—a human conveyor belt from road to crest. But since her wage for an eight-hour day was only eight *annas* (14¢), a modern mechanical conveyor would have done it less cheaply.

"Why are they forced to do such labor?" I asked Aloysius. "Have they no men to support them?"

"They are very much married," he grinned. "The women of this particular tribe are allowed five husbands each, there being so many men where they come from. Their men are those you see packing heavy loads into the mountains and back. While they are away, these women work at whatever they can—in the fields, on the roads. The rest of the time they devote to raising their children."

Suddenly Aloysius motioned to the rickshaw boys to turn down a sinuous pathway into a ravine.

"I have something to show Madame," he confided. "The water-babies."

A moment later I knew what he meant. There, at the bottom of the glen, lying side by side in a row in the shade of giant *Deodar* cedars, were perhaps a dozen babies. They were sound asleep, yet nowhere was there sign of a guardian.

I looked at my guide questioningly.

He bent low and whispered, "They belong to the women we saw working by the road. The little ones are brought here every day."

Placed close to the steep bank they lay on small beds of rushes covered with thick gunny sacks and blankets.

"They leave them here like this all day?" I asked.

"Yes, Madame."

"Who takes care of them while the mothers are at work?"

"No one, Madame. They are left absolutely alone."

"You mean to tell me they lie peacefully asleep all day—just like that? What happens when one of them wakes up and decides to crawl off?"

"That never happens." The man beckoned me closer and pointed to the nearest infant.

It lay there a picture of bliss, its small head cushioned in a wreath of reeds. But then I saw that the head was wet, that a trickle of water was falling from above on to the child's forehead. It was icy cold. A curved piece of tin had been stuck in the bank so as to catch the seepage from a spring and funnel it, drop by drop, to the forehead. Similarly, each tiny sleeper had been placed beneath its own tin trough, and each head was covered with the bright liquid beads.

"It is the constant dripping that keeps them asleep," Aloysius explained. "But that alone would not be enough. Look!" He drew my attention to a scar in the center of a baby's forehead. "That is where the mother blistered the skin with some vegetable caustic to make it sensitive and hurt a little. The cool water soothes the burn and lulls the child into a deep sleep, broken only when its mother takes it up."

I was dumfounded. "How did anyone ever think up such a thing?" I questioned.

"It is a very old custom," he replied. "Some say it began at the time of the Mohammedan conquest to prevent the children's crying from disclosing the hiding places of the Hindus.

"Others believe in an older tale—the story of a king who forbade his daughter, Bibi, to marry her lover because he was poor. In spite of this they stole away together, and before long a son was born to them.

"Hearing of this, the king ordered the child put to death. But it could not be found. The two lovers had vanished—mingling with the common people and working in the fields so that no one knew them. The king would not be cheated, however. He instructed his soldiers to kill every male child in the land.

"The people fled and hid in the woods with their little ones. The soldiers, led by the children's cries, followed, sparing none. Bibi and her husband were safe for a time, for they had fled deeper into the forest than the others—though it was only a matter of a few hours before the king's men would find them, also.

"Knowing they would be discovered unless she could quiet her child, Bibi prayed to Naini, the water goddess, for help. A spring suddenly gushed forth from the hillside above them and a thin stream of water was directed on the baby's head, putting it instantly to sleep."

"Bibi's son was saved, then?"

"Yes, Madame. And since that time the same trick has saved the hillwomen of India a great deal of trouble with their young ones."

I recounted the story to Barnum when he returned—tired, but radiant—from Bilaspur.

"Stuff!" he commented, briefly. "We'll have to chuck the fairy tales now and get back to bones. Pilgrim gave me a lot of ideas on new places to dig."

## 15
### The Respectable Murderers

After breaking camp on our return to Siswan it was the open road for us, each week some new place—Dhok Pathan, Kalka, Chukwal, Chandigarh, Chenji—until Barnum had prospected all of the Punjab and most of the Punjab States; also sundry territories belonging to His Highness the Maharaja of Patiala.

Scouting was the order of the day. We travelled light, stopping in dak Bungalows along the route. We seldom used the tents, and then only for a night. Abdul, relieved of his culinary duties, served as Barnum's assistant. His helper, Fasil, took over as full-time cook—a grave mistake. We didn't mind his straining the soup through the long end of his turban, holding the bread between his toes while he made toast or sneaking "snifters" on the side. But when he calmly tailored himself a pair of Sunday pantaloons out of my checkered dishcloths, I fired him on the spot.

A cook was excess baggage, anyway. The daks came equipped with complete corps of servants—chefs, *khitmagars* (waiters), even *punkah-wallahs*, small boys who sat on the porch manipulating huge fans tied to their big toes

The Indian dak, or rest-house, is one of the world's great institutions, ranking with the American tourist court. Few villages, however remote, cannot boast of their daks—each a small oasis of comfort in a land where hotels are scarce and comforts rare. Built by the government out of cool brick or stone and maintained at state expense, they vary in size and the number of accommodations. Some, near the large centers, are more like country inns than bungalows, with several suites of large furnished rooms opening on a central lounge. Others are less pretentious. Wherever we found a dak we could be assured of clean restful quarters for a trifling fee. One rupee (thirty cents) a day engaged a suite of bedroom, dressing-room and bath, the latter equipped with the Oriental luxuries of tin foot-tub and commode. Meals were extra, of course, and it was good policy to tip the help.

As daks rented on a first-come-first-served basis, Abdul went ahead to make all necessary arrangements beforehand. Only once, outside the small town of Dhok Pathan in the Punjab, did he fail to have a fresh clean dak ready for us to step into. That was because it was doubling as the community court-house.

Two native-brothers had just been tried by the village elders for murdering the postman. Motive: fifty rupees and a pair of shoes. The courtroom was emptying as Barnum and I rode up, and from the broad smiles on the faces of the prisoners it was evident that they had either been acquitted or given extremely light sentences. Led down the road, the pair broke into loud and lusty song, several bystanders exuberantly joining in.

All was not as it seemed, however, according to Abdul. Actually the brothers had just been condemned to death by hanging. True, the case would still have to be reviewed by an English judge in a higher court. Such was the law in the British-controlled Punjab to prevent any miscarriage of justice. But there was slight doubt that the verdict would be approved and the culprits speedily executed.

"Then why the rejoicing?" we wanted to know.

"These men are *loot-wallahs*. They belong to the robber caste," Abdul informed us, going on to explain the psychology of the Indian criminal mind. Burglary was an honorable profession among members of this caste; an inherited way of life. Just as the priest was born to pray and the warrior to fight, so the *loot-wallah* was born to steal. That was his karma—his destiny—the will of God who had seen the need for plunder and pillage in the human family as well as in the animal kingdom.

They had been good respectable thieves, these two brothers, stout fellows in the crusade against organized society. They had shared their loot with their comrades, worshipped regularly at the shrine to the outlaw-god, observed the customs traditional with their kind, one of which forbade any male member of the band to marry before he had committed the required number of larcenies. But no one had expected murder of them. They didn't come from cutthroat stock. Petty pilfering—nothing higher—had been the family business for centuries. Now, overnight, the ambitious pair had soared from mediocrity to renown! They were heroes. They would die martyrs to their cause. Remembering this, they sang.

We did some looting of our own around Dhok Pathan; several weeks of it, in fact. Successive raids on a half-dozen isolated graves paid off handsomely in skulls of extinct species, all nicely petrified. They included those of three antelopes, a horse and two mastodons—one with a tusk seven and a half feet long. Ghouls Brown and Abdul did the dirty work while their "moll" guarded the swag in the hideaway dak.

The small jobs we could handle alone but carrying off the mastodon skulls, weighing close to a ton apiece, required outside aid. So we hired a few of the village stonemasons, known as *mistris* in this part of the world, for the heavier chisel work. They did their job well, thanks to Abdul.

Then came the problem of transporting the big heads to the railroad at Chukwal, sixty miles distant. It was the first four miles that had us worried—four miles of tortuous rock-strewn badland that cut off the dig from the nearest road. Camels could make it, but their load-limit was only eight hundred pounds, and Barnum naturally refused to break the skulls up. Bullock carts could pack the weight, but the terrain would have them stopped in fifty yards. We were stymied.

Called in for consultation, Abdul considered. Then Abdul disappeared, returning with a gang of twenty-odd coolies borrowed from the nearby oilfield of Khoar. They were equipped with enough picks and shovels to push a road through to the marooned fossils in no time at all.

A jubilant Barnum hauled out our dynamite and blasted a passage for the road crew. The result wasn't a highway, exactly, but it served for the oxcarts. Two of them, drawn by eight bullocks, took eleven days to make the sixty miles from dig to railhead.

Having checked the haul in at Chukwal, we swung southeast over the Diljubba Ruka Pass and prospected the broken country around Hasnot. . . . Nothing was ever better named. Our luck picked up again at Chenji with a rhino skull and several jaws from the red beds there.

So it went with us—dak to dak—bone to bone. Except for an incomplete camel skeleton, together with some odds and ends of elephant and hippopotami collected in the Chandigarh foothills, our specimens were running definitely to heads. This tickled Barnum pink. It was usually the other way 'round. And when a skull of the rare *Sivatherium* fell to us, his joy knew no bounds. Next to the Siswan turtle, it proved our prize catch.

The creature must have been something of a prehistoric oddity. It sported two pairs of massive horns, a proboscis, the teeth of a giraffe, and a body that might well have resulted from an affair between an elephant and a rhinoceros. The experts give it membership in the giraffe family. Leave it to my man to find the freaks; he wasn't named Barnum for nothing.

With this last acquisition, I counted up our gains. They made an impressive list; impressive enough, I thought, to have earned us a vacation—a real honeymoon in the high cool land of Kashmir's shining mountains and lotus-covered lakes. Kashmir! Could any honeymoon spot sound more romantic? There was a more practical side too. The monsoon season was approaching and the Punjab beginning to bake. Every white man in the country was fleeing to the hills for the summer months. Every white man, it seemed, but mine. Barnum merely shrugged his shoulders and prepared to sweat it out. His reason: he'd made previous engagements with more fossils—as yet undiscovered.

I finally persuaded him to cancel some of these appointments in favor of the Kashmir trip. So everything was settled at last. I planned to go on ahead, hunt up a houseboat, get in supplies, and make ready to shove off on a tour of the lakes the instant he arrived.

"Promise to join me *before* the month is up?" I coaxed.

"One month it is, Pixie," he replied, firmly. Then he softened: "But after that,—"

"After that—the honeymoon!"

# 16
## I Ride the Royal Mail

From Rawalpindi on the plains to Srinagar in the Vale of Kashmir, the wise traveller goes by private automobile. I know that now. I didn't then. I rode the Royal Mail as super-cargo, and it was the most hair-raising experience I've ever had for the price of a bus ticket.

At some time in the remote past my conveyance had been a truck. Currently, it was the closest thing to a fossil on wheels that I have seen. Certainly of prehistoric vintage, the contraption should have been gracing the hall of some mechanical museum, instead of jeopardizing the lives of innocent tourists. Of such accessories as doors, windows or windshield, nothing was left. All that remained was mere skeleton—a chassis, steering-gear, four wheels and something under the hood that sounded like the Pittsburgh steel mills but did little to get us anywhere. It rode as if it had been petrified for several million eons. Just to step inside the thing was foolhardy, much less sit it out for two hundred miles over the front ranges of the Himalaya Mountains on a road that put any roller-coaster in the tunnel-of-love class.

And whom should I pick for a driver but a frustrated speed-demon! Not that he drove fast. He simply seemed to—the only man in my long career of motoring who could make twenty miles an hour feel like two hundred. I nicknamed him "Speedy."

Speedy was a nice enough chap—on the ground; a very considerate affable Sikh with a wonderful black beard that tucked under his chin. But as soon as he got behind the wheel, something snapped inside him and, presto!, he became another Barney Oldfield racing to glory down the last lap of the five-hundred-mile Indianapolis Speedway classic. His dual-personality had me worried from the start.

I sat up front with him, the sacked mail riding in back under lock and key. On our right front fender perched a human claxon— a small boy with a rubber bicycle-horn strapped to his shoulder. This sounded continuously as we drove out of Rawalpindi. Near the outskirts a flock of geese, flying north to the Kashmir lakes, heard our honkings, veered and "buzzed" us.

All at once the horn stopped. My eyes shifted to the right fender. The honker was gone.

Startled, I questioned Speedy.

"Oh, him falling off," came the casual reply.

I half-rose to my feet in the jouncing vehicle. "*Fell off?* Then in God's name, man, turn around!"

Speedy looked mildly surprised. "Why?" he asked. "Not needing him now. No more traffic. He always falling off at city limits."

Such an arrangement was right in line with everything else that happened on that screwball trip.

On leaving Rawalpindi, the road started to climb up out of the hot Punjab plains into the mountains. From 2000 feet it went to 6000, the car's radiator perking like a coffee pot.

With altitude, Speedy grew lighter in the head and heavier in the accelerator foot and my blood-pressure took a turn for the worse. By the time we reached the crest of Murree Hill—somewhere between 7000 feet and the Hereafter—he was Jimmy Doolittle piloting a B29 over Nagasaki. I wasn't worried now. I was scared stiff.

"Let me out! Let me out!" I yelled frantically as we approached the brink of a precipice, the bridge connecting with the other side a slender hair suspended above a fathomless canyon. I piled out and closed my eyes as the truck crept across ahead of me. Then, taking my heart in my mouth, I followed on hands and knees. The darn thing didn't even have *sides*.

From there to Kohala in the Jhelum River gorge twenty-five miles distant, we dropped five thousand feet. The old crate performed beautifully while dropping and, with a tail wind, really made time. To this day, I claim we did a somersault on this stretch—though nobody will believe me. Down the hills and through villages we'd roar with a volley of backfires, scattering natives and chickens to right and left. A twist—a turn—then like a shot along the ragged edge of nothing, with never a thought for our necks or the brakes! Like all Hindus, Speedy believed in predestination. As for brakes, they were something to stay away from—like "likker" and bad women.

"Never touch brake," he would say proudly. I learned the reason why. The only way he could make the next hill was on the momentum gained by speeding full-tilt down the last. Except for cows, he slowed for nothing, not even human beings. Cows were sacred.

Once one of these holy bovines cut our momentum enough to stall us on an upgrade. "No spark," came the diagnosis. The next three hours were given over to tinkering. A crowd gathered. Eventually it formed a pusher-brigade that shoved us to the top. Then, with the help of gravity, we coasted the rest of the way to Kohala.

No sooner had I stepped from the truck than a bumptious little creature, smelling strongly of perfumed soap, placed himself squarely in my path and, without a word of introduction, proceeded to pepper me with questions.

"Name?" he sputtered, flourishing pad and pencil. And, before I could answer, "Destination?"

"Srinagar," I stated coolly.

Shorty scribbled diligently, looking up from time to time as if taking my description. Suddenly his pencil shot out and pointed directly at my face. "Occupation! Quick . . . occupation! Answer quick, please," he demanded.

That had me stumped and, with the pencil continuing to make passes at my nose I couldn't think very well. I finally managed to blurt— "Wife."

The pencil halted in mid-air. "Wife!" exclaimed the voice behind it. "You say you are wife? Where is husband?" A wife unaccompanied by a husband was evidently something new to his experience.

"Mr. Brown is in Rawalpindi," I snapped back, nettled.

This elicited a broad "Ah-h-h," followed by, "Will madame please coming with me?"

What could I lose? Speedy had to send back to "Pindi" for a new magneto, so I was stuck in Kohala for the night, anyway. Furthermore, the man had canceled my bus ticket and my luggage was in the hands of his hefty accomplice.

Bags following, we made our way up a slope to a small dak bungalow. Inside, the cross-examination continued. How long had I been in India? Why was I going to Kashmir? Why was I travelling alone? And always, at an unexpected moment, he would whirl on his heel with, "Where is husband?" My time wasn't altogether lost, however. Between questions I managed to dash off a few cards, and a letter to Barnum.

Then, in a surprise move, he dismissed his assistant, locked the door and pulled a flask from his pocket. "Madame liking drink? Nice Haig and Haig whiskies." He poured himself a peg. "Making feel good," he added, insinuatingly.

Patience crumbled. *"No!"* I shouted. "And you had better open that door at once or—or the American Consul will hear of it."

The fellow's chubby face paled. "No oping door. You oping suitcase—all suitcase." He flashed something that looked like a badge, and I thought it wise to comply.

Expertly, his fingers flew through the contents, feeling linings, testing for false compartments, fluffing my silks for concealed jewels or contraband currency. Next he sampled the small bottle of liqueur and shook up the tin of American coffee, my prized possession.

He looked up quickly. "American lady ver' pretty," I heard him say.

That did it!

No doubt my highly combustible temper comes from being about as Irish as is possible for a native-born American. There was a handy water chatti on the stand behind me. I aimed for the head. He ducked. The pitch went wild. The stand followed, crashing into the opposite wall and knocking Rama, the seventh incarnation of

Vishnu, out of his frame. What happened after that is a bit hazy. When things cooled off, I found myself in an adjoining room with the door bolted on my side and Passion-flower rattling the knob on his side, whining through the keyhole— "Please, madame . . . oping, please. Me showing credentials."

I let him rave a while. His voice filled with pathos. "Please to hear, madame. Me good mans. Me police-inspector—doing duties."

So that was it! I opened the door, lit myself a cigarette, and listened to his whining explanations. Unescorted females, it seemed, were always grounds for suspicion in India, especially along the Rawalpindi-Srinagar highway in the spring; and more especially at Kohala, the border-station between the British Punjab and the native state of Kashmir. Spring was the migrating season for "fancy ladies" and, according to police records, the Primrose Path led straight through that particular village to the greener pastures of the Vale. Traffic had been so heavy that year, and illegal entry so frequent, that the Kashmir girls had risen in arms, the politicians were working and the Punjab Police had put their ace-detective, my plump friend, on the job.

The instant I boarded the truck that morning, a spotter had rushed the news to headquarters, and Inspector X, pride of the force, was ordered to intercept the white-slaver smuggling herself into Kashmir in a mail sack. Now that it was all a huge mistake, there were apologies, lamentations and laughs.

The Inspector, he assured me, did not like his job, despite the fact that he was the envy of the entire constabulary. "Me married man," he reflected sadly. Still he had to admit that the work had its pleasant side. One met so many interesting people!

"Well, c'est la vie. One must live," I said. . . . We parted the best of friends.

Next morning I was back in the cab with Speedy as he maneuvered up the valley of the Jhelum. On our left tumbled the foaming river, fresh from the glaciers of the high Himalayas. On the right rose pine-covered slopes, ridge on ridge to the distant peaks of the Pir Panjal. Tiny mud shacks sprawled on the mountain-sides amid green patches of wheat. They marked the towns. We had the road practically to ourselves. Occasionally a *tonga* flashed past—light, two-wheeled gadabouts drawn by fast horses—or a heavy *ekka* loaded with luggage, servants or freight.

At scattered points long trains of double-decked bullock carts had stopped along the roadside. The great beasts that hauled them were down—beautiful spotted creatures reclining under the wagon-tongues—some wearing gunny throw-overs to keep off the flies; others with gilded horns, necklets of bells and bright blue beads. The wagon-beds bulged with freight. In the upper stories, under tent-shaped covers, the drivers lay asleep. Only at night, when motor traffic was prohibited, were the bullock caravans permitted to travel the highway. At dusk the long lines of carts would pull out from their encampments on another leg of the three to four weeks' journey from lakes to plains.

Presently we met the down-mail and both drivers pulled off the road to pass the time of day. Two hours of it, to be exact! Conveniently an Englishman was aboard the other truck. Comparing notes, I soon learned that the mail-line's two-day schedule for the Rawalpindi-Srinagar run was nothing but a boast. "Four days is the usual time for the trip," he stated. "I know. Been riding these beastly buggies for years."

Twenty miles from Kohala lay Domel—and more trouble. At the edge of town a rickety toll-gate slammed down in front of us—a huge wooden beam split in the middle and wrapped with wire where one of Speedy's buddies had gone through three years before. Some characters with guns swaggered out of a custom's shack, examined our papers, looked over the cargo and removed a gasoline tin from under the mail bags. Next, they were removing Speedy himself and hustling him off to the local jail. The Law again. Ho hum!

Around a table before the toll station sat several oldish men with pencils in their turbans, the sure sign of officialdom. One of them—obviously the Commandant—sat on the table paring his toenails with a long knife.

I addressed him point-blank: "On what charge are you holding my driver?"

His reply was a grunted, "Smuggle."

Heavens! More of that nonsense. "I thought that misunderstanding had been cleared up in Kohala with the Punjab officer," I said. "I'm travelling to Srinagar of my own free will, Mister, and not as contraband. What's more, I intend to get there."

Twinkletoes didn't know what I was talking about. My driver was being jailed for smuggling extra petrol over the border without a permit, he said. That's what was in the gasoline tin.

How would the driver go about securing a permit?

"He could buying one," the man ventured, "but . . ."

The "but" trailed off in a vigorous rubbing of his big toe.

"But what?"

The tough face managed a woebegone expression. "He having nothing to buy it with."

Money! I should have known. "How much?" I asked.

"Five rupees."

Speedy was released and we left Domel in a cloud of noxious backfires.

Another dak that night in Garhi, a cluster of mud shanties in a valley of wheat where wild lilies grew. Up at dawn!

"Finish troubles today, Memsahib. Srinagar tonight," Speedy promised.

"Lunch in Rampur?" I questioned hopefully, noting a place on the map midway between Garhi and Srinagar.

A nod.

"Hold your hats, boys!" I yelled back to the mail bags. "We're barging through today."

We didn't have lunch in Rampur. We didn't have lunch. The truck developed a new set of ailments, stopping dead in its tracks every quarter-hour or so and requiring more of Speedy's "fixing." This consisted of transferring small quantities of water from the radiator to the fuel tank, then blowing into the tank till exhausted. It worked! With a series of terrific explosions the car would leap forward as if shot from a cannon—only to stall again a short distance farther on.

Once while my companion was busy tapping the radiator I sneaked a stick into the gas tank. Although it showed less than an inch of fuel, Speedy refused to add more. "Plenty petrol, Memsahib. Plenty petrol."

Recollecting the five-rupee fine back at Domel, it added up to a neat bit of "chiseling." My driver was "stretching" the gas to sell in Srinagar—at a good profit, of course.

Came meal time. Rampur—miles ahead. I was famished At this point my nose detected a questionable odor in the cab. At first I thought it might be cheese, then Speedy, then something dead. I dismissed it as none of my business. But later, as the day waxed warmer, the smell *demanded* attention. The vapors seemed to be coming from under my seat. Investigation then disclosed a long slim form wrapped in dirty newspaper. I kicked it. It was stiff. A breeze whipped over a corner of the paper and there staring up at me was an ancient dried fish. Anyhow, here was food.

Speedy explained that he had bought it in "Pindi." He always carried a little snack on his trips—just in case. I could have kissed him.

Instead, I kissed the fish. He took half, gave me the other. Nothing ever tasted more scrumptious. We ate head and all. We even licked the bones.

At Rampur we made a stop, but solely for gas—straight, unadulterated motor fuel which I purchased for the Royal Mail with the understanding that it was not to be watered along the way. What remained in the tank when we pulled into Srinagar was to go to the driver.

As a result, the truck's temperamental behavior miraculously improved. We made the home-stretch non-stop, through Baramula where the Jhelum valley widened into the lush beginnings of the Vale, down the straight level avenue between rows of giant poplars which lined the last thirty-five miles. They stood out like sentinels, sharp and black, as darkness fell over the swampy meadows and rice fields behind. And far beyond, in all its 26,600 feet of majesty, rose snow-crowned Nanga Parbat of the Himalayas, eighth highest mountain in the world.

It was night when the first lights of a town twinkled through the trees. We drove through a sleepy bazaar, reached a bridge that crossed a sleepy river. Moonbeams played hide-and-seek among the houses along the shore. Small covered boats hugged the banks.

A voice at my side said, "Here finish troubles, little Memsahib."

Srinagar! Mile-high capital of Kashmir—City of the Sun—journey's end! So soon?

Speedy looked at me and smiled. I couldn't.

## 17
### HIGH-PRESSURED IN KASHMIR

My first day in Srinagar started off with a near-riot. All dolled up, I was in the act of making a dignified exit from the hotel lobby, when a band of noisy hawkers ambushed me. They were trying to sell Kashmir. Apparently they went on the theory that once they obtained a foot in your mouth they could sell you anything. All I saw at the moment were beards, hands, mouths, going like mad. "Lady Sahib wanting cook, horse, turquoise, tailor, basket, bearer?" Falling back on the old refrain, "Only looking, not buying," I dodged a shawl. A woodcarver popped up from behind and almost clipped an ear off with his sample. Finally a turban went flying through the air—minus a head.

Now, a turban is the most important part of an Indian's attire. For him to be suddenly separated from one is much like depriving a Britisher of his pants. In short, he feels completely naked. With a bellow, the lidless merchant wheeled and shoved back, setting off a round of shoving. Somebody got mad and yanked the nearest beard. A shriek!—and the fight was on.

When the tears of laughter cleared from my eyes, a little mouse of a man had parked himself right under my nose, taking full advantage of the fracas.

"Madame wanting nice houseboat?" he squeaked.

I practically fell into his arms. "You're my man!" I exclaimed. And so, while the lions fought, the mouse ran off with the prize, leading me to his waiting *tonga*.

Down the muddy lanes we clopped, through a labyrinth of narrow crooked streets teeming with humans—mostly Moslems, men, women and children—all busy as bees. A happy-go-lucky lot, compared with the people of the plains! The men "yo-ho" as off to work they go; the women sing as they prepare the meals, their wooden pounders beating a tattoo in the rice crocks. Baksheesh bums trot alongside the carriage, shouting something that begins with the "Bow-wows," and ends with "Cheap Mama," if nothing is forthcoming.

In Tailors' Row, a small boy sits in a doorway fanning the fire under a pressing-iron, two henna-bearded sheiks cutting cloth in the background. Others are grouped around a Singer hand-machine trimming stacks of the coarse wool "Numdah" rugs we often see at home. In a tangle of colored yarns and silks, skull-capped men embroider scarves and fancy garments.

Not a shred of their art shows in the clothes on their backs. They don't give a hoot about the "new" look; styles were set some three hundred years ago by the Emperor Akbar. When he invaded Kashmir in 1626 the inhabitants already had been rummaged by so many conquerors they had nothing left but their shirts, and these he took without a struggle.

Akbar, a doughty warrior, liked resistance in his enemies. When the weak Kashmiris failed to put up a good fight, he was understandably disappointed. Here was the fairest land in all the world—and no one to defend it. Were there no men in the Vale? Apparently not!

So the decree went out that both men and women should dress alike. The outfit selected was—and still is—a one-piece Happy-Hooligan nightie with a hole cut through for the head, and wide loose sleeves. As it happened, this costume turned out to be the most convenient all-around garment ever designed. God knows what they have up their sleeves, and they could get away with murder under their skirts. In fact, from the "expectant" look of the men, the overlapping of sex appeared to have gone too far. Then I learned that they carried charcoal heaters under their clothes to keep them warm on chilly mornings.

Despite the drab garb, there is a certain rakish beauty about the Kashmiri gals. Slim, sharp-featured, all they need are "the red silk stockings and the green perfume."

Of course, you never really see the *high-caste* women. They clink along the street à la Ku Klux Klan, disguised in billowy white *bourkhas* with tiny squares of lace across the eyes to see through. If they're dare-devils they may go around with their noses nude; and should you accidentally surprise a group of them with faces stark naked, they will pick up their skirts and throw them over their heads. Then—git, Brother, git!

Absorbed by these sights, sounds and activities, suddenly I became aware that it was taking an awfully long time to arrive at our destination.

I addressed my mouse-like guide with an anxious, "Is your boat moored on the Jhelum?"

"Oh yes, Lady Sahib."

"Is it a big boat?"

"Oh yes, Lady Sahib. Ver' big!"

"I don't want too big a boat."

"Oh no, Lady Sahib. Is not too big."

Over an old wooden bridge we rattled, into the Harpie Bazaar where "Flowers of Delight" on painted pillows lured men through devious passages to high balconies. More dark alleys. We seemed to be moving further and further *away* from the waterfront.

"Where in heaven's name *is* this boat?" I demanded at last.

"Soon—soon."

A lurch—and he drew up before a funny little shop, stepped down and motioned me in.

"What's this? Where is it, the boat? I thought you were taking me to your boat!" I said sharply, retaining my seat.

A look of amazement spread over the chap's face. "Boat, Lady Sahib?" he quavered innocently. "What boat?" His hands shot into the air. "Il-Allah . . . no boat. This my *shop*. Me cheap okay silk and woolen weaver.

"You're also an okay weaver of the truth!" I exploded.

The fellow salaamed as if acknowledging a high compliment. "Telling truth not good business," he smirked. "You not coming my shop first. Me partner with Suffering Moses, manufacturers of superior papier-mâché makers."

Thus, my first encounter with that redoubtable breed, the Kashmiri trader, whose slogan is, "Any means to a sale!" And though their business methods are unethical, they do prove effective—as demonstrated by the armful of candlesticks, bowls, boxes and gimcracks with which I returned to the hotel.

Only one thing was missing. I still didn't have my houseboat.

## 18
### All but the Groom

Apparently, if I was ever to locate one I would have to dodge the merchants and find a man who had nothing to sell but his services. Hamid was that man—a cabby recommended by the hotel. He piloted a *shikara*, a water-taxi to you. Armed with a list of addresses from the travel agency, Hamid and I sallied forth.

"Sunwarbarg," I directed, settling back among the cushions. The panorama that slid past was like nothing I had ever seen before. "The Venice of the East," the guide book called it. With good reason. Streets were waterways and the populace lived afloat.

The Jhelum River is the Grand Canal. For two miles along its banks the city stretches—a fantastic jumble of palaces, hovels, mosques, bazaars. Down the middle passes the river traffic— *shikaras* in gay trappings darting between huge grain barges; scows weighed down with produce, and thatch for houses; passenger- *dungas* with matting-covered roofs; gobs of moving color in flower boats bound for shops along the bank. Cries of boatmen float across the water, mingling with the hubbub of native life. The "all clear" signal sounds as, with the flash of half a hundred paddles, the Maharaja's State *Parinda* glides by—long and sleek—a dream of dazzling white and gold.

Luxurious houseboats ride at anchor in the shade of huge *chenar* trees that mark the Bund, the city's busy esplanade. Past the European quarter the river flows—the English Church, the Club, the Residency, Post Office, Parsi shops, Punjab Bank, Cockburn's Agency.

Farther on you blink and stare at the tall white pillars of the Royal Palace; the Golden Temple where the Prince of Kashmir worshipped; the Harem with its lofty balconies and mysterious latticed windows through which the favorites watched the passing show. On the opposite bank, bedecked with flags and colored banners, are the palatial houseboats of the Ministers of State.

At the tail end of all this magnificence sprawls the city of Almost-but-not-quite. The native quarter, that is, where the buildings are on the point of collapse but haven't quite hit the ground. Wooden skyscrapers angle several stories in the air, propping one another up in a haphazard sort of way. Housekeeping here is a perilous task. Frequently a goat is visible gazing serenely down from an Upper-story window, or nibbling the grass that grows on the roof. Washing takes place on the doorstep where the women also fill their water-pots.

Seven bridges span the Grand Canal. If you can slip by the third without being lured on the rocks you are lucky. And it isn't the sweet song of the Lorelei, either. It's a hail of "Salaam Huzur's— Howdy Your Highness" from hoarse-throated hawkers who all but fall out of their shop windows waving pots and whatnot at you. In case you can't hear, they have signs out. For example, there is "Sidik Joo, Plain and Worked Silversmith"— "Rahmana T, OK Best Woodcarver"—big black letters proclaim "Habib Shaw, Finest Rug and Shawl Merchant," with shawls so fine you can pass them through a wedding ring. These are but a few of the shops on the river that sell the handicrafts for which the Vale is famous.

Out of the Jhelum we skimmed and into a side canal walled with high levy-like embankments. Under a bridge, shadows. Then sunlight again. The banks widened. Our *shikara* burst into an expanse of broad shimmering pond. Sunwarbarg!

Along the wooded shores nestled innumerable houseboats, from mere floating shacks to ornate residences. A few dips of the paddle drew us up to a huge gingerbread affair with carved-wood porticoes and spacious observation deck running from stem to stern.

Hamid read my thoughts. "Owner wanting plenty moneys," he said. "Four hundred rupees for month. Inside ver' good—having okay piano. Colonel M. Sahib living here last summer. Plenty parties."

"How many rooms?"

"Six rooms for livings, three for bathings."

"Too large."

We paddled farther along "Society Row," past the "Boat of the Seven Gables," with its pert dormer windows peering out from the shingled roof. Another sported a veritable roof-garden where some tweedy Englishmen in deck-chairs, sipping gin 'n bitters, eyed me through the potted plants.

The *shikara* headed toward a cozy little cottage-boat off by itself under a drooping willow. I caught a glimpse of freshly-painted sides gleaming white—of ruffled curtains—of railing'd roof-deck mellow in the glow of a bright awning with fringe around the edge.

"Darling!" I enthused, jumping aboard.

Inside was cute as a bug's ear—just enough Kashmiri atmosphere in the flowered rugs, shaded lamps, and drapes on the long French windows. Behind the divan, a soft India print fell from ceiling to floor. There was a dining-room all golden in the sunlight, and through the open door I saw the boudoir. It had a gay bedspread, and walnut chests, one for Barnum, one for me. How heavenly to spread ourselves around after months-on-end in a suitcase! There was a pantry, too, ready to he put to good use. And wonder of wonders—a bath! What closed the deal, however, was the fireplace.

Here at last was my dream come true—our honeymoon-boat.

Our private *shikara* waited at the rail, red cushions and all. Tied astern was the cook-boat, and from under the matting cover the boatman and his family were breaking their necks to have a "dekko" at the new boss.

We went into conference. The boatman, it seemed, would cook for twenty-five rupees a month; his son, Hussan, would be bearer for fifteen. Then, we would have a waterman and a sweeper for fifteen each. Add one hundred for food, fuel, and the like; one hundred fifty for rent, and there you were. "All this, and Heaven, too" for not much over three hundred rupees a month—one hundred American dollars!

I couldn't wait to break the glad tidings to Barnum. "Note our new 'at home' address, Dear One," I wrote. "Houseboat No. 6, Sunwarbarg, Vale of Kashmir, India." Such a honeymoon would make up for everything!

Life on the river was sweet idleness. It drifted from day to day and, like the river, one hardly knew its passing. I would awaken to the singing of birds—bulbuls, orioles, thrushes, wrens—to the saucy chatter of the starlings and bold mynahs who beat you to your toast if you didn't get there in a hurry. Many mornings I breakfasted on deck as the slow mists rose from the margs, or clearing, the sun and hot above the mirrored world of shore. There would come the splash of paddles—the mirror rippling—sharp-prowed *dungas* nosing out from the banks—the water-people off on their business of the day.

The sun rose higher. And with it appeared long slim rowboats laden with flowers and vegetables from the floating gardens of Dal and Nishat—tomatoes, cucumbers, squash, small eggplant to cook in so many ways; or lotus roots and water-chestnuts, should you have a taste for them; and strawberries, melons and delicious fruits in season—all fresh and ripe and paddled to your very door.

Small wonder these aquatic people think of themselves as descendants of Noah. From cradle to grave, life to them is a watery affair lived in diminutive replicas of the Ark. Everything floats, even the land.

Because there is a great scarcity of terra firma in the region of the capital, that fact has furnished the Kashmiri a way of turning an honest *anna*, the Indian two-cent piece. Resourceful farmers have rigged up floating gardens made of long strips of reed held in place by poles driven into the lake bed. They pile mounds of earth on these, then seed the earth. . . . You'd be surprised what springs up practically overnight. Now and again during a storm, one of these floating gardens slips its moorings, running amuck, so to speak, and you're apt to find someone's vegetable patch on your poop deck in the morning.

I spent the sun-drenched days, far into the nights, poring over books about the "Happy Valley"; catching up on my diary so sadly neglected since those well-remembered words, "I thee wed" . . . anything, everything, to hurry through the hours till Barnum came. Oh, the fixin' and the doin', the marketing, the preparations for his arrival, perusing maps and planning an odyssey through the lakes and up the Jhelum, scouting out places to stay, to fish— "and taking mulberries for bait," Hussan would smilingly remind me.

I even planned the homecoming dinner. Our goose would hang high with applesauce and trimmin's just like Christmas. Of course, I might change my mind and have squab, chicken, duck, or mutton. "But no beef," cook would say with a long face. "This Hindu State. Punishings ver' bad for killing bulls. In old times boiling butcher in oil."

Today if anyone accidentally runs down a calf or any member of the herd he gets off with a mere seven years imprisonment.

But, in all this planning I was only marking time till my husband arrived. He might appear any day now. I wrote descriptive letters about the scenery, the quiet nooks reeking with romance. Weepy letters. Passionate letters. "Here is the Shangri La of Lovers," I would tell him—and always I closed with the words of the poet Moore, "Remember Love, the Feast of Roses." That's how sentimental I felt.

June would see the Vale abloom. I existed through that month, watching the roses bloom and fade. When would he come? Would he *ever* come? I almost died of waiting.

Whenever I felt too lonely there was the boatman and his family. They would do anything for me, and likewise do me for anything—but who cared? They were so generous of their time, and little surprises, too. No work was too long.

There was Hussan the oldest son, my guide, shopper and faithful shadow; the two younger boys who helped around the boat— "*chotah-wallahs*," I called them, "little fellers." The good wife was my Kashmiri Madonna, always with babe in arms or at her breast.

When the last light waned, from deep in my easy chair on the upper deck I could see them snuggled together in the cook-boat, Hussan stretched out astern strumming a *sitar*—the Indian version of a guitar.

Then the loneliness would overwhelm me. "Hussan, pack my things. Tomorrow I leave to join the Sahib," I'd bravely announce. So each morning found the luggage neatly set out on the bank.

On one occasion, happening to lift one of the suitcases, I thought it suspiciously light. I opened it. *Empty!*—like all the others.

"Hussan, what is the meaning of this?" I demanded.

"Well, Memsahib, to speak truth, after first five orders packing seeming waste of time. You never leaving anyway," was the candid comeback. The-departure-of-Lady-Sahib-in-the-morning had become a standing joke among the crew—a mere formality when she had the blues.

The best thing about houseboating is the ease with which you move from place to place without bestirring yourself. A few words to the skipper and off you float, house and all à la magic carpet.

"The lakes!" I ordered one day. To the chant of punting coolies, we swung from the bank and were poled out-stream. We idled along through the warm sunny day, the air sweet-scented with blossoms and visions of loveliness on all sides—fields of yellow mustard flower, meadows of poppies and lilies, sad patches of purple iris covering the silent bosom of the dead. We passed pink and white cherry trees, soft and fluffy like Milady's powderpuff, and airy huts of natives peeping through clumps of poplars, lookouts on their roofs to spot the bears that might invade their cornfields. Mingling colors everywhere, from the opalescent depths to the dazzling snows of the mountains!

At night we tied up beneath some big *chenar*. Huge trees, these, as much as sixty feet around. Many are several hundred years old, their hollow trunks cut into doors with locks—the homes of Holy Men. They furnished the final touch to fairyland.

Now and then I struck out alone with Hussan in the *shikara*. Small, light, it could slither through the lotus pads and browse along the shore. How many trance-like hours I spent in the haunts of the Great Moguls whose reigns made up the Golden Age of India! Akbar, the father, who conquered Hindustan, amassing wealth untold; Jehangir, his playboy son, who squandered it on pleasure-gardens, magnificent buildings and beautiful women; Grandson

Shah Jehan, warrior, statesman, artist, lover, who left the world the architectural jewel, the Taj Mahal.

I paddled the waterways in the wake of their imperial barges, through masses of pink lotus into Nishat, the Garden of Delight, Jehangir's rendezvous with Nur Mahal, his "Light of the Palace." Another unforgettable day I spent at Shalimar the Abode of Love, amid almond blossoms, sparkling fountains and carpets of brilliant flowers—dreaming my way through marble pavilions where lovely ladies played, and Shah Jehan whiled away his summers. This was the very garden of the Kashmiri Song, "Pale hands I loved beside the Shalimar." My hands were pale beside the Shalimar, but who was there to hold them?? In all this loveliness there must be love!

"But," my fossil-hunting husband wrote to ask, "is there any bone?" Had I seen anything that looked like a "skinkus" in the fair Vale? The letter was postmarked "Chukwal." I had a notion to fire right back, "*No!* but I know where there's a 'skunkus' in a place called 'Chuck-all'!"

An idea popped into my head. Why not try the old Kashmiri technique on him? So I wrote that there were "skinks" by the bushel, and that the banks of the Jhelum were crawling with bones. He didn't bite!

I guessed I hadn't measured up to his expectations as a fossil-huntress, anyway. Here, in the beauty around me, I had found my heart again—temporarily lost in the dust of ages. He could blimey well *stay* down on the hot plains; I was taking a trip up the Jhelum. Incidentally, I might have a look around for some bones. Just *maybe*.

The Jhelum is a river flowing out of India's past, its upper reaches lined with the ruins of age-old cities still retaining, in their sad decay, some of the magnificence of former glory. There stood Avantipur the ancient capital, a striking mixture of old and new, with hovels built among the remnants of Buddhist temples.

Farther into the land of long-ago the river wound. At Islamabad I moored the boat and rode overland past the well of the sacred fish fed by Hindu Holy Men—on—on—to the Temple of the Sun.

Perched on a high plateau, it is the largest temple in Kashmir, its architecture strongly Indo-Grecian. The place was named Martand.

Ghost-cities all, these places are peopled by the thousand-year-old shades of Kashmir's early kings—the mighty Laladitya who built the Temple of the Sun; Avanti, the Good, who filled his halls with treasures; Sankara, the Bad, who looted them; the wicked Queen Didda who closed the Hindu reign, and Mahmud who brought Moslem rule to India.

Their ruins still stand and the Jhelum still flows southward to the sea and conquering hordes still come and go. Nothing disturbs the ageless sleep of India. Perhaps *now* is the time of her awakening.

When I returned to Srinagar letters galore awaited me but no husband. I took comfort in the thought that surely he'd be here in lotus-blossom time. Well, the lotus finally bloomed, and while it flowered my honeymoon hopes were fading. I dashed to Wular Lake, swam its waters said to cover a bygone city of some four thousand years ago, rowed out to the island supposed to have been built on one of its temples.

Back to Srinagar again—and a bee-line to the post office. At last a crumpled note Barnum scratched, apparently, on the fly. "I love you," it said, "I love you—*but . . .*"

Another day passed. Two. Then a telegram in the mail! So he was coming at last. Weak with excitement I tore it open and read: "Kashmir trip off. Leaving for Baluchistan. Meet me in 'Pindi.'"

I sat quite motionless for some minutes. Then I got a wire off,
"Can't wait to see you. Wonderful news about Baluchistan. Returning 'Pindi' immediately."

So it was goodbye to dreams of love beside the Shalimar to the music of zithers and the poems of Laurence Hope. They ended in a grand rush for more bones in more Godforsaken country on the wild-west frontier of India.

19

THE NAWAB REGRETS

"It's a Baluchitherium we're after," said Barnum, waving a cablegram from America— "worth ten honeymoons any day."

"Never heard of the beast," I retorted.

"Well, you've missed a heap of animal. Next to a dinosaur, a Baluchitherium is just about the biggest thing in fossils. Alive, it was a giant hornless rhinoceros; stood eighteen feet tall at the shoulders and had a skull four feet long. That makes it the largest land mammal on record.

"President Osborn of the Museum wants us to round up a reasonably complete skeleton of the creature," he breezed ahead. "I figure the best place to start in is where the original discovery was made—in the Bugti Hills of Baluchistan."

Now, the Bugti Hills were border-country and, as such, in the usual state of political combustion. Normal relations among the trigger-happy populace consisted of "fussin' and feudin'" and worse, with Bugti tribesmen shooting up brother Bugs and Waziris hunting Waziris. And when the feudin' fell off there were always a few Pathans or Afridis around to pep things up with a small-scale war.

That the atmosphere was hardly congenial to the peaceful pursuit of bones we were made thoroughly aware in the border town of Jacocabad. Well-meaning Britishers assured us it would be hard-sledging in the Bugtis. They didn't advise an expedition; we were much too young to die.

They regaled us with horrendous tales of white men who went into the hills never to be seen again—probable victims of the "skin game." This was an extremely popular sport among Afridi women. They bound their prisoner in a tight net, sharpened up their knives, and gently relieved him of all skin protruding through the meshes. They then shuffled the poor poor fellow up for a new deal, and so on till he was completely skinned alive. Ants saved funeral expenses. While we listened, our kindly hosts bound us with several lengths of official red tape, just in case the stories failed to discourage us.

But they didn't know Barnum. Opposition only steeled his determination to have a Baluchitherium if it killed him. He talked. He wheedled. He humored. He spoke of the forward march of science, of British sporting instinct—and finally he had his way. Official barriers gave, and the matter was referred to the native ruler, the Nawab of the Bugtis. If my husband could secure *his* approval the Imperial blessing would follow.

Only one difficulty loomed. The Nawab was attending a state festival at the time—a *durbar* at Sibi. That meant indefinite delay before our petition even reached the royal ears.

"Why wait?" Barnum reasoned. "Sibi is less than a hundred miles north. I'll strike while the *durbar* is hot, and catch the old boy in a festive mood." So saying, he boarded the train for Sibi.

He came back greatly elated over the result of his trip. The Nawab, after consulting his official astrologer, had finally granted full permission to prospect his lands. There was but one proviso: we must take a military escort with us for protection.

While dickering for the bodyguard, a terse communiqué came through from Sibi. His Nibs, the Nawab, had changed his mind. The expedition was off. Upon conferring with his Finance Minister, the Bugti chief had decided that Americans were too expensive. The last one, an oil geologist killed by raiding Pathans inside his borders, had cost him a twenty thousand dollar indemnity paid the widow.

Barnum appealed once again to the British authorities, using every argument he could muster.

"Sorry, old top," they said, in effect. "It's the Nawab's territory. Cawn't be overruled, you know."

Heart-heavy, we cabled our bad news to the Museum. The reply took our breaths away. We were to drop all plans for the Baluchistan venture, finish up in India, and make arrangements for an expedition into the interior of Burma.

Burma! In exactly three hours we were on the train to Rawalpindi, Baluchistan a memory. We tarried in "Pindi" no longer than it took us to wind up the unfinished business connected with our Punjab work, and ship the heavy equipment off to Calcutta.

The Museum did acquire a Baluchitherium—though not from Baluchistan. Some months later, Roy Chapman Andrews, then in Mongolia with an expedition, was to discover the skull and part of a skeleton of this gigantic beast in the Gobi Desert.

One thing that delayed us slightly in getting away from "Pindi" was Abdul's uncertainty as to whether he would accompany us to Burma. We hated the thought of getting along without the big, efficient fellow yet didn't want to urge him against his better judgment. He finally told us that family affairs made it inadvisable for him to travel so far from home.

He saw us off at the railway station, and I shall never forget our parting. As the train pulled slowly from the platform, we waved a last goodbye. I could see him standing there leaning against a pillar, and, as I watched, the big man appeared to wilt—to shrink. His head was turned slightly away and he was making faces in a struggle to keep back the tears. Suddenly his hand clutched wildly for the loose end of his turban, raised it to his eyes and covered them. I knew he was sobbing.

I cried too. Our wonderful jinii, our giant—he seemed so much more like a lost little boy now. What would we do without him?

"Abdul Azziz . . ." Barnum lingered over the name. "We'll never meet his like again."

We never have.

# 20
## Dinner with the Maharaja

We broke the trip to Calcutta by a stop-over at Patiala, where Barnum was to call on the Maharaja and present his geologic report on the lands he had surveyed for him.

A black Rolls Royce met us at the depot, and whisked us to the palace. There the royal secretary, an Englishman, received us and escorted us, not to the audience chamber, but to a broad turf playing field ringed by spectators in turbans, army khaki and white duck—all intent on watching a cricket game. Clearing a place for us on the sidelines, two dusky attendants brought wicker chairs and tall drinks, with the help of which we puzzled through a game of English baseball.

The secretary pointed to one of the players a head taller than the others. "The Maharaja," he explained. "After Polo, cricket is his favorite sport. He captains the Patiala team, you know; doing a fine job of it, too. We're leading Benares State by a comfortable margin."

The man beckoned to a British couple some distance away, and turned to us again. "His Highness has been looking forward to meeting you and learning something of your fascinating work. He hopes you will stay for the costume ball tonight."

Barnum and I looked at each other in amazement. We hadn't included costume balls in our bone-digging itinerary.

By that time the British couple had strolled up. They proved to be the Maharaja's head-gardener and his wife, the George Burrows. It was to them, after the secretary had gone, that we confided the sorry state of our wardrobe.

Said Mrs. Burrows, "We'll fix that right now. George's man can take care of Mr. Brown's outfit, and my tailor is a perfect genius at dreaming up costumes on the spur of the moment."

If ever there was a fairy-godmother this charming English lady filled the bill. And I was an eager Cinderella. Like magic she spirited away my drab travelling suit, producing in its place a creation in red, white and blue inspired, she said, by the American flag.

My Prince Charming was the Sheik of Araby—Barnum in flowing robes, a lascivious leer, and a jeweled dagger in his waistband.

As the two of us stood in the palace anteroom that night waiting to be presented, we had the eerie impression of having suddenly dropped out of the Present onto the doorstep of some medieval prince. So far as we were concerned, an air of unreality hung over the hail. Beyond the vaulted doorway stretched the ballroom—white and gold and endlessly long. We saw guests gathered in tight little knots on the floor—bright patches of color—and behind the hum of voices came strains of music, weird and exotic, strange to Western ears.

Through rifts in the shifting crowd we caught an occasional glimpse of the Maharaja himself. Even at that distance there was something majestic about him,—a superb *aloneness*, as of a mountain rising from a plain. About him wheeled his courtiers in prescribed orbits like planets circling a central sun—and then a passing cloud of chiffon or tulle or cloth of gold would close him off from our view again.

The guests were presented to His Highness in small groups, singly, or in couples. Giddy with apprehension, I watched and listened, awaiting our turn.

The moment arrived. We heard our name called in clear stentorian tones: "Mr. and Mrs. Barnum Brown of New York City." I slipped my hand in Barnum's arm and we sailed out onto the polished floor past a sea of faces until we came before the Prince.

He was rather plainly dressed in the costume of a Scotch Highlander. By his side, however, a tall fair Scot doubled for His Highness in the royal robes and jewels.

My husband bowed. I curtsied. The Maharaja bent forward. "Welcome to Patiala," he said, adding with mock severity, "I had hoped you would visit me sooner."

We moved to one side then, making way for the next guest. It was over. I breathed a sigh of relief. Barnum squeezed my hand.

Dinner was an event. Preceded by His Highness, the guests flowed into the adjoining banquet hail, three sides of which contained long white tables banked with flowers and set with crystal and gold that glittered in the light of tall candelabra.

An official seated Barnum and me on the Maharaja's left. Behind the guests, along the walls uniformed butlers moved silent and swift, anticipating every wish.

The kitchens poured forth their goodness: salvers of filet de Pomfret; platters piled with roast mutton or great slabs of Yorkshire Ham; more salvers bearing unctious arrays of chutneys and Indian condiments; roast breast of guinea hen; steaming heaps of rice pilaf thick with nuts and candied fruits. There were white wines, red wines, champagnes and sweetmeats of every description. The clink of crystal blended with the tinkle of laughing voices.

His Highness was easy to talk to. Jovial, without hint of condescension, he had the gift—rare among Hindus—of a sense of humor. Often his booming laugh filled the hall, followed by a ripple of answering titters down the tables. Our talk covered considerable territory—Barnum's work—the projected trip to Burma—America. But mostly we talked of the Maharaja.

He was a Sikh, member of that illustrious race of fighting men in whom the British found such able allies in the administration of India. Although he lived as a medieval prince in a world all his own, politically, at least, this man was very much of a modern. Immediate dominion status for India was his dream, each state to share in the government of the federated whole. Later, while serving eight and a half years as Chancellor of the Chamber of Princes, he was to bend all his efforts in that direction and prove himself one of the world's foremost champions of States Rights.

In the course of the dinner the conversation drifted around to his queens—the maharanis. There were six of them, one for every day of the week except Sunday. "And I am looking for a seventh," the Maharaja appended, a twinkle in his eyes. Barnum coughed and nudged me.

Divining our thoughts, His Highness directed our gaze to the gold-latticed screening at one end of the hall. "My queens have been with us all evening—seeing but unseen," he said. "You may be sure they have examined you quite thoroughly by now, Mrs. Brown. As an American, you are of great interest to them. They have heard much about the freedom of women in your country, and, though they admire you for that freedom, they do not envy you. It is the nature of Indian women to cherish security above personal liberty. Some day they may desire equality with their menfolk, but to give it them now would be like turning house-canaries loose in the jungle."

While he spoke, a hand, reaching from behind, slipped something beside my plate. I looked down. It was a music-box—small, round, delicately painted, with a tiny Dresden china couple twirling around the top to the strains of the Emperor Waltz.

"A little memento of your visit to Patiala," His Highness whispered.

Following dinner, the Maharaja and Barnum, accompanied by several Ministers of State, disappeared through a great door. This left me with His Highness's "stand-in"—the bejeweled Scot.

According to his own conservative estimate, there was a cool million in jewels on his person—part of the seven million dollar hoard usually kept in the royal vaults. A fringe of the famous Patiala black pearls edged the large white turban. Collars and ropes of diamonds hung from his shoulders in scintillating profusion ending in a center pendant big as a robin's egg. Pearls pale and milky showed in a band across his waist. Over the heart blazed a magnificent diamond star.

I was spellbound, and I told him so.

"Ai! and 'twill be the death of me yet," the Scot answered with a short laugh. "Might be stabbed in the back any minute, toting these bloomin' stones around."

"You're plainly a man to be trusted, I'd say."

"'Tisn't that, M'Lady. His Highness wouldn't much care if I did walk off with them. He thinks more of his British military decorations than all these baubles."

Somehow I could believe that. Sir Bhupindre Singh was that kind of man.

The Scot and I strolled out to the terrace. The night air was cool and a bright moon lit up the palace grounds. Through the shadows I could see the lights and dark shapes of many buildings—each in turn explained by my companion. He pointed out the six palaces of the queens adjoining the main palace, the *Moti Mahal*; the heavily-guarded children's palace where the royal offspring were being reared by European tutors and governesses; the Imperial garage with its fleet of eighty-seven cars. Other structures were lost in darkness.

Whenever I think back upon Patiala, it's always with a feeling of unreality—as if that night had never really happened but was only dreamed. I can still see the towers and hear the music, but memory holds nothing clear or sharp. It's like some exotic tale of long-ago with which my mother might have sent me to sleep.

Only the music-box tells me that the dream was real. Something has happened to its mechanism. The little figurines are still. No longer do they dance, nor does the music play. It was a fragile thing, meant for gentler times than these. But somehow—broken, silent—its link with the past is even stronger, now that I know my memory of Patiala is of the actual India that was.

The world of the maharajas has come to an end. A lot of evil has gone with it—but beauty, too, and courtliness and honor.

PART TWO

# BURMA

"*The thoughts of his heart,*
*these are the wealth of a man.*"
—The Buddha

21

A TOUCH OF TEXAS

Safari in Burma!

For two months Barnum and I thought of little else. We planned for it by day, dreamed of it by night. It was a song in our hearts, a constant refrain—safari, safari, safari. India seemed a world away and a thousand years ago; we now lived in the future. What wonders lay ahead—what new mysteries to explore? The sooner we started, the sooner we'd know.

We promptly fell in love with the gay Burmese. Our first weeks in their happy country passed like a day, with only fleeting glimpses of people and places snatched *en passant*—Rangoon, blurred and bustling; night train to Prome; river-steamer up the Irrawaddy, talking "shop" with the skipper—hoping, planning, weaving visions to the music of the lazy river's passing.

We left the boat at a place called Yenangyaung, meaning "the creek of the smelling waters," and aptly named. Everywhere we inhaled the thick vapors of crude petroleum, while oily residue lined the banks.

Three miles east of the river, in a region more desert than jungle, rose a forest of derricks, most of which belonged to the British-controlled Burmah Oil Company. Barnum, always interested in anything in the ground, whether solid or liquid, wouldn't think of missing the opportunity to see this largest of Burmese oil fields, and one of the oldest in the world.

"What would my colleagues back at the petroleum institute say!" he exclaimed.

One of the field supervisors, who acted as our host, was Mr. T. S.—a long, thin man browned by years of tropic sun, and speaking with a drawl unmistakably Texan.

"Mighty happy to meet you folks," he greeted us. And he *was*. No doubt about that. From the way he wrung our arms off you'd have thought we were a couple of old amigos from down San Antonio way.

He had his men fix up for us a little bamboo bungalow on high stilt legs, with lattice walls and grass-mat rugs. So we settled down to the life of Riley for seven glorious days—and nights.

Our evenings were spent at the local club, the "American-Yen," where an almond-eyed *chef du bar* in a pink silk bathing cap (the Burmese *baung-gaung*), silk skirt and white linen jacket, skimmed about serving iced drinks in carved silver bowls. Lime squash, beer, highballs, and a local firewater made with palm sap, called a "toddy collins,"—these headed the list. Dinner was home-cooked, with dill pickles, berry pie, and that nectar of the gods, American coffee. We were really living!

After coffee came the cheroots—dainty morsels a foot long and almost two inches thick passed around in a black and gold lacquer box. The best of them contained tobacco, spices, shavings of tree bark and a dash of opium. The "whackin' white cheroots" of Kipling fame are the corn-huskers of the poor people.

Most of the personnel at the field were Americans, and it was wonderful to be among our own kind again. Far into the morning, or as long as the cheroots held out, we would sit around talking, catching up on the news and swapping snappy stories. I think the night we had the most fun was at a so-called "moonlight picnic." Three married couples and two jolly bachelors made up the party. Palms, pagodas and Oriental music we had in abundance. Everything was complete for a moonlight picnic. Everything that is, except the moon. There wasn't a scrap of moon.

Our days we spent amid the dusty bustle of the drilling crews smelling the crude, feeling the movement and excitement, hearing the rattle and bang of drill tools, the pound of the pumps, the incessant "queek-quawk" of the walking-beams and the whirr of winches paying out cable to the hoarse cries of drillers. These men were from Oklahoma, Texas and California and had signed up for a three-year hitch on the "Burma shift." Their talk made me homesick.

With our host, T. S., guiding his big Buick among the derricks, we listened as he pointed out the various features of the field.

"The company has put up twelve hundred modern rigs on this property," he said. "They give us well over two million barrels of oil a year. Get our best production from between three and four thousand feet down."

Barnum remarked how close the wells were crowded together, noting that many of their legs actually overlapped.

T. S. smiled. "Yep. Reckon we have more drill holes in a given area than any other field in the world. Comes from the old system of parceling out land here. Back in the days when the Burmese kings granted the rights to drill this section, they made property allotments so close together that not more than forty-five feet separated the wells. The British increased this to sixty feet when they came, but they're still packed together tight as sardines.

"You see," he continued, "this is a very old field. They say it's been operating for at least seven hundred years. The diary of a Chinaman who visited 'Yen' in the late thirteenth century states that it was going full-blast then."

The car halted before an aged wooden structure being worked by natives. "This is one of the old hand-dug wells," T. S. explained. "There are sixty of them here—operating, that is. They're called 'twinzas.'"

"'Twinzas?'"

"Comes from the title given the original holders of the land grants. The term means 'one who obtains a livelihood by possessing an oil well.'"

We watched as brown-skinned Burmese hustled about the derrick. Nearly four hundred feet below, at the bottom of the hole, T. S. told us, a native diver bailed fresh crude from the oil sand into a five-gallon gasoline tin. He was naked except for a g-string and a diving helmet connected by hose to the surface. Here two men, members of his family, kept him supplied with air from an oxygen pump.

"Three barrels a day is a pretty good average for a well like this," said T. S. "The diver stays down from one to three hours at a time, depending on conditions, and works by the light reflected to him by that mirror over there." He pointed to a looking-glass canted at an angle over the well opening, and went on. "Years ago, before they used diving helmets, a man could remain in the pit only a half-minute—digging like mad and trying to breathe as little gas as possible. It took him half an hour to recuperate. Twenty descents a day was his limit."

At that instant a shout from one of the workers sent a group of coolie girls into action. We had noticed them before, standing single file beside the well. Now, tug-o'-war fashion, each grasped a section of rope that led over a grooved pulley down into the shaft and pulled—hauling to the surface a spindly little diver, covered from head to toe with dark heavy oil.

"Hard work for two rupees a day," the Texan drawled. "That's only about sixty-five cents our money, you know. The men at the pumps get three rupees, and the rope girls about eight annas—fifteen cents. Quite a gap between that and the forty rupees our men stash away in a day."

"How long does it take them to dig one of these wells?" I asked.

"For a four-hundred-footer? About two years. Takes us less than a week with our modern equipment. Of course, not all their holes are that deep. Most run between two and three hundred feet. That's where the uppermost oil zone is located.

"A small hand-hoe serves for digging," he continued, seeing our genuine interest, "and, as the well deepens, they board up the sides. When a hard sandstone layer is encountered, they shatter it with a heavy, pointed iron weight. This they suspend by rope from a wooden beam. It hangs there, pointed down directly over the center of the hole. When they cut the rope, down she goes—striking the bottom with enough impact to smash the hardest rock. A man is then lowered, and the weight retrieved with a rope."

"Effective, no doubt . . . but primitive," observed Barnum.

"It shor is primitive, Mr. Brown," the supervisor agreed. "These methods were obsolete two thousand years ago. D'y' know that the Chinese dug brine wells three thousand feet deep with bronze bits and bamboo *before the time of Christ?*"

"How do they go about selecting their well sites without any knowledge of geology?" my husband asked, after a pause.

T. S. laughed. "They use the 'doodlebug' method—the 'doodle-bug' in this case being a small stone image of an elephant. They place this little image on a flat rock in the sun, arrange a circle of gifts around it, and sit down to watch."

"Watch for what?"

"Well, as the sun drops the elephant's shadow lengthens. The offering first touched by the shadow is carefully marked, and in this direction, a mystic number of steps from the image, the natives dig." Our Texan smiled wryly. "It's not any more fantastic than the way we used a divining rod to locate oil in the States a few years back."

Barnum agreed. Science had come a long way from superstition in a very short time.

A week at Yenangyaung, and we were off again—chunkin' up broad old Irrawaddy. We had servants with us now—a Madrasi couple named Mari and Dos whom a kind American lady, their employer, insisted on accompanying us.

Who knew what awaited—what joy and what tragedy—as we pushed steadily northward into the land of the Smiling Buddha!

## 22
### CREAKING THROUGH BURMA

Pakokku! At long last we had reached the little Burmese Village that was to be the starting-point of our expedition. Before us stretched the long-awaited jungle-trek, adventure ... perhaps even a honeymoon. Suddenly the Future became Today!

I could see Barnum from the cottage window—Barnum no longer the distinguished gentleman-scientist in spotless white duck, or masquerading in flowing robes as the Sheik of Araby. He was all-explorer again—shorts, loose shirt and topi—directing the loading of our wagon train.

To the unimaginative, ours may not have looked like a very impressive outfit—four bullocks, two small matting-covered carts, a pair of sway-back saddle horses. But had we the mounts of Genghis Khan himself, with a hundred of his camels, together with an escort of Royal Bengal Lancers, I couldn't have felt more excited. They travel best who travel light—and slow.

The villagers, too, were agog as they clustered around the caravan. Eyes popped and heads wagged at our strange equipment, the curious risking furtive peeks inside the carts. Even the tiny brown kewpies riding their mothers' hips were in high spirits, gurgling incessantly and reaching pudgy hands toward the white faces. Never had the community seen such activity, and I knew there were some in the town who would not be altogether happy about our leaving.

The hour of departure arrived. Then the actual moment when the bullock drivers nudged the oxen with their long poles. Thus the great Burma Expedition creaked into motion. I looked at my watch. 7 A. M.

"Take it away!" I shouted to Barnum at the head of the column. He turned in his saddle and threw me a salute. We were off into the Unknown, our world-on-wheels between us. In his pocket rested our passport to the jungle—a letter signed and sealed by the Resident Commissioner to all village headmen along our route, requesting them to furnish us supplies as needed.

My husband led the way astride an ancient nag while I brought up the rear on an equally antiquated mare, gentled by age. Mari and Dos traveled in the carts, jammed in with our supplies.

The natives lined the road as we filed out of town. Many followed to the very edge of the forest, waving cheery farewells . . . and then we were on our own. Destination: Monywa, nearly two hundred miles distant through thick teak-bamboo jungle. Object: rainbow-chasing, the rainbow being an elusive ribbon of varicolored rock—red, blue and yellow—that unwound through the woodlands, alternately appearing and disappearing beneath the heavy undergrowth. Somewhere along this rainbow we would find our prehistoric pot-o'-gold, we hoped.

Life on the road began before dawn with the crackle and flare of the campfire, and Man, like the Witch of Endor, brewing a spot of tea. Fortified with a hasty draught while loading up, we were on the move by four o'clock. It was here that I learned the gentle art of riding horseback in my sleep.

Speed was no object. This was fortunate, since our pace was set by the plodding bullocks—two miles an hour—thirteen miles a day. At any rate, it allowed plenty of time for Barnum's off-trail sorties in pursuit of bone.

There was slight chance of his losing contact with us because of our musical cart-wheels. Often he could hear them miles away. They kept him constantly informed of our whereabouts.

Squeaky carts, I might say, are the glory of the Burmese bull-whacker, the mark of his profession. Never greased, the noisier the axle the greater the glory. He takes professional pride in the

distinctive pitch of his wheels, and is said to be proficient in distinguishing his fellow workers by theirs. In the same way, he can
detect the approach of a stranger by the sound of an unfamiliar
hub. A few good rattles also help maintain one's standing in the
Burmese teamsters' union.

Thus we had music while we trekked, whether we liked it or
not. Personally, I liked it. A merry melody, harmonizing with the
sounds of the jungle, it invariably sent crazy rhythms running
through my mind.

The season of the year may have helped a little too. Spring is a
glorious time to be in Burma. It has all the color and spice of our
northern autumns. The woodlands blaze with rich yellows, reds
and russets, and each passing wind sends fresh cascades of leaves
tumbling to the ground. Beneath our feet the spongy foliage of the
teaks, their great branches merging with the sky! Nostalgic odors,
as of October, rise from thickets of aromatic shrubs to mingle with
the fragrance of flowering trees and the heavy perfume of orchids.

The forest is alive with birds—wild canaries, bluejays, screech-
ing parakeets, saucy "jungle crows" with long black tails and henna
breasts. Wild fowl scuttle through the underbrush; bright-colored
cocks, drab hens. Deep in the shade of scarlet *dak* trees the snow-
white heron nests beside the jungle pool. Through the long hot day
and far into the night the poor-will whistles to his mate. Such is
the gay, sunlit façade of the Burma forest.

But in the gloom far back from the trail, amid the tangled
growth of creepers, climbers and barbed vines, exists the grimmer
side of jungle life. Here armies of huge red ants are devouring the
carcass of a python; there a pretty green reed turns into the deadly
Russels Viper. Distant trumpeting of wild elephants cuts through
the wilderness. Everywhere death in the arms of life!

And over it all, come the hysterical "whoop-poo whoop-poo-o-o"
of the gibbons, like the cry of an ailing child. Seldom were we without a noisy band of these small apes keeping tabs on us, watching
our every move with their white-rimmed "spectacled" eyes. Whenever we did anything they took exception to, they let us know about

it in no uncertain terms—showering us with pods and twigs, or just showering us.

Golden maize stacked in the trees and a handful of bamboo huts beneath feathery tamarinds, marked the villages. About nine bells, before the heat of day, we would creak up to our happy jungle home—a shed with palm-fringed roof and all outdoors for walls. Known as *zihats*, these shelters are built by the charitable for pilgrims and weary travellers like ourselves.

Each community has its own pagoda, with a large assortment of Buddhas, and bells forever tinkling in the wind. All through the hours these bells peal out their melody, constant reminders of the great philosopher, Gautama Buddha, whose teachings guide the daily lives of the Burmese. On the outskirts of town there is usually a rambling teak monastery where the monkhood lead quiet lives of study and contemplation, educating the youth in return for the villagers' support.

Civil affairs are in the hands of a headman, the *tedje*, who settles disputes, offers advice and, in general, looks after the welfare of his people. They regard him more as a father than a public official, and under his guidance live like one large family, content with food, smokes and entertainment for the day.

The people themselves are poor but happy. There are no great fortunes among them; wealth is evenly divided. A man gains nothing by keeping things for himself, they believe. All value lies in giving. If he has more than he needs to cover his simple everyday wants, the Burman digs a well for the community, builds a bridge, pagoda or school. He never hoards. He follows pretty closely the precept of the Buddha: "The thoughts of his heart, these are the wealth of a man."

But for all his healthy outlook on life, there is great need in the back country for simple drugs of all kinds, especially antiseptics.

Barnum had done well in stocking our outfit to serve as a sort of travelling medical dispensary, in addition to all else we carried. At every village we distributed liberal supplies of quinine, aspirin, iodine and potassium permanganate. The permanganate, diluted with water, proved most effective in checking the eye disease so prevalent in the Orient.

At each village, when the inhabitants learned of the free pills and eye-wash to be had for the asking, they came crowding around our outfit like children around a Good Humor ice cream wagon.

23

NAT TROUBLE

We were jogging merrily along through a particularly jungly stretch of country one morning, our little covered-wagon train singing its squeaky song to the open road. Night mists were rising from the paddy fields up through the tall still palms. The hollows were deep in shadow. Here and there, sunbeams broke through the wall of vegetation, sprinkling our path with bright sequins of yellow. From somewhere came the sound of a barking deer, the flutter of frightened wild fowl. Happy, happy land! I thought.

Musing, I could see Barnum riding far ahead. Soon I watched him disappear around a bend in the trail. As we approached the turn, the slender trunk of a young palm somehow became caught under the lead cart, bringing the party to an abrupt halt. All attempts to back the cart off proved futile; the tree remained firmly wedged between the wagon's bed and right wheel.

Jabbering excitedly, our two drivers promptly fell to discussing many elaborate schemes, with the usual Oriental ingenuity for making mountains out of mole hills. Not one of their plans contemplated extricating the vehicle. They were all aimed at—*rescuing the tree.*

After considerable palaver, the drivers commenced a diligent, if somewhat unorganized, wrestling with the wheel—one pushing, the other pulling.

Suspecting the worst, but wishing to give them the benefit of the doubt, I rode up and inquired what they were doing.

"Taking off wheel," came the bland reply.

"What?" I demanded. "Going to all that trouble when one good swipe of your *dha* (knife) would cut the tree down?"

The two natives shrank back and regarded me with shocked surprise, if not downright hostility. Blurted the chap with the *dha*, "No! Memsahib. No! No! cutting down tree, nats having no home. Nats making plenty troubles."

Nats? What jibberish was this? Who or what in heaven's name were nats and what did they have to do with the situation, anyhow? I was about to dismiss the matter with a hearty "Nuts to Nats!" and order the tree felled forthwith. But, thinking better of it, I calmed down and turned to my servant for an explanation. "Dos, what's this nat-business they're talking about?"

Dos shifted uneasily, scratched some non-existent bite under his turban, and pondered the treetops. "Well, Madame," he said finally, carefully choosing his words for such an important explanation, "Burmans having ver' strong beliefs in spirits by name nats. Nats making houses inside trees. Chopping down palm same as chopping down nat's home. Nat get ver' mad, follow and put bad hauntings on us, pushing carts in hole, making Sahib lose way." His voice dropped. "Maybe even killing."

The tree was left standing, even though it did take the better part of an hour to remove the wheel and replace it again. I didn't want any blithe spirit putting a hex on us and worrying the daylights out of our boys.

Barnum was hopping-mad when we caught up with him at the next fork in the trail. "'Bout time you showed up," he snapped. "Can't you see there's a storm brewing? What caused the delay, anyhow?"

"Just a little nat trouble down the line," I answered, casually.

He eyed me suspiciously, and unconsciously massaged behind his right ear. Then, "Hummph," he grunted, and let it go at that.

Our brush with the palm-tree nat was but the first of a series of delightful encounters with these creatures. Who, for example, should we run into at the very next village but His Celestial Highness, the Rain Nat!

As Barnum had feared, the black clouds we had noticed earlier maneuvered into position directly above us and burst. By the time we reached the village—a darling place called Gyat and about as big as its name—we were drenched to the skin.

But so was everyone else. Every last inhabitant of that tiny community was out in the downpour making whoopee—shouting, laughing, running about like mad, and, general, having a time for himself.

Our first move was to look up the local *tedje*, acquaint him with our arrival and secure lodging—waterproof, if possible. But the *tedje*, a pudgy soul who looked like an overgrown cherub, wearing a benign smile and little else, was a leap ahead of us. He had been informed long ago of our coming; the jungle "grapevine" had seen to that. As for accommodations, a *zihat* had been ready and waiting days for us to step into and make ourselves at home.

As we sloshed toward the cottage I questioned him about the wild goings-on in the village.

"Making feast to Rain Nat," he explained. "Need much rain here. We giving thanks."

"Rather early for rain, isn't it?" Barnum put in.

"Oh yes, Sahib. Rainy season not come for month yet—mebbe two. We making special plea to Rain Nat for come bring rain before time. Nat hear. Right away make rain." The headman waved a hand toward the sodden landscape.

We agreed. Rev. Rain-maker had certainly done right by lil' ol' Gyat. Jupiter Pluvius couldn't have performed better himself.

It came about that the *tedje*, being administrator of the village, had many dealings with the nats. As a matter of fact, only because of his "connections" with them, he informed us frankly, did he hold office at all. Be that as it may, he had a vast fund of information concerning the creatures, information of which I availed myself at every opportunity. As a result it was the nats who taught me much about the Burmese people and their character.

One has only to see through the childlike eyes of the Burman, and all Nature comes alive with pixies, elves, nymphs and gnomes. Collectively known as nats, these creatures are inextricably woven into the folklore and religion of the country. Every hill, tree, field, river, rock and garden, not to mention the Elements, has its own particular nat proprietor—and the jungles are veritable fairylands.

Such out-and-out nature-worship is never sanctioned, of course, by the Buddhist monkhood. But knowing that any attempt to uproot these primitive fancies would be folly, they wisely tolerate it. For my part—being Halloween-born of an Irish mother who early saw to my schooling in pixie-lore—I was completely captivated by this living Oriental mythology. I only hoped that we might happen on some likely Leprechaun who would lead Barnum to a rich cache of Burma's prehistoric bones.

There are two general classes of nats—good and bad. Naturally, it is the former whom the people strive to cultivate, but they cannot afford to snub, much less openly offend, the evil spirits. For bad nats are spiteful demons who can raise more hob than a truckload of our more civilized gremlins. They have been known to set an entire village afire—just for the hell of it!

Hence, it behooves mere Burmese mortals to keep in the good graces of their nat neighbors at all times. With this in view, believe it or not, nat-houses are provided by the villagers for the accommodation of resident and tourist flats. These are small frame affairs, much like bird-houses, placed atop tall posts along roadways, in gardens, beside dwellings, where any nat—good or bad—may obtain food and lodging *gratis* for as long as he cares to stay. Some of the more pretentious are of fine workmanship, displaying ornately-carved wood-work and filigree trim, walls aglitter with bits of mosaic looking glass, and fairy-size porches—lattice-screened—set with miniature potted plants.

Each morning before any member of a household dares touch a smitch of food, a substantial breakfast must be served the spirits. An offering is placed within the tiny shelter, its interior tidied up—and heaven help the family should some fussy nat, feeling out of sorts, drop in for a bite, only to find that the ants and birds have beaten him to it.

In return for being "kept," the spirit-folk are supposed to forego their mischief-making and behave. If a nat does occasionally run amuck, it's purely for want of something better to do.

As far as I was concerned, Gyat's Rain Nat was of the bad variety. True, he loved the thirsty pastures, but he likewise drove from their homes all creeping, crawling, slimy things whose mission in the world was to terrorize a small memsahib, meaning me.

One morning my boot contained a "ball of yarn" which materialized into a hairy tarantula the size of my hand. There were poisonous black and red ants, scorpions that nested in the provision box, and, for diversion, foot-long orange-colored centipedes. Worst of all was the tiny black and white mosquito whose sting brought the dread jungle fever. In the smothering humidity they bred by the millions.

Without Flit or DDT, the best defense against these insect hordes was a strong "isolation" policy, resulting in a special sleep set-up each night when I crawled into my cot, squeezing myself as small as possible so as not to pull the cot in after me. All my worldly possessions went in, too—six pieces of clothing, two of which were stockings; trousers and shirt to make up my headstone, while interior decorating was done with topi and boots.

Following which, my maid would query, "Madame gotting legs in cans of kerosene?"

"Yes, Mari. Legs of cots all set."

"Madame gotting mosquito net pulled tight?"

"Yes, Mari. Madame tight."

On what, God only knew!

## 24
## "Ver` Bad Womans"

Barnum had extensive bone-hunting scheduled for the country back of Gyat just as soon as the town's water sprite decided to turn off unseasonal showers. Meanwhile he labeled fossils, caught up on his notes and worried about the weather, while I pottered around making our new home cozy—which wasn't too difficult.

The *zihat* was on the small side, about twenty feet by twelve, thatch roof, rough teak floors. But a little fixing brought about amazing improvements. We even managed to solve the space problem with some ingeniously-hung gunny-sack "partitions," converting the one large room into a neat three-room bungalow—parlor, bedroom and bath. Downstairs, in quarters of their own, our servants, Mari and Dos, made themselves comfortable. What a pair!

For color, the ace of spades had nothing on Mari who, in turn, had very little on herself—so far as concerned clothing. A tight jacket fastened beneath her breasts; a white *sari* draped her limbs; the in-between was nude. A study in ebony and ivory, Mari! She would have bartered half her charms "to be so white as Madame," and Madame had tried, without success, to swing a hip like Mari.

Her high falsetto in the kitchen below was what usually awakened me in the morning. More often than not she was bawling someone out. "Burmese mans got ver' bad eye," her talk would go. "Make spoil the bread. Can't fry the water. Always looking, looking" . . . climaxed by the thud of a skillet on a bare back. This was Mari's normal way of starting off each day. It cleared the air and got things moving, she contended.

When Dos was within shooting distance she took her morning mood out on him, often assuming the role of militant suffragette as she read the riot act to her even blacker half. Dos combined a wild mixture of Hindustani, English and Burmese, mostly the latter which boasts the most voluptuous swear words. He took it all good-naturedly for a while. But when the pots and pans began to fly, he retreated to me for protection.

Backing out of the line of fire, I would call to Mari, "What seems to be the trouble?"

A wail in reply. "Damn da Dos! Always rounding with Burmese girls and making lies. If telling truth, thinking stupid."

But Mari had another side. Sometimes, in the languid afternoons, I would stretch out on the matting floor of our little grass shack for a moment's rest, my maid beside me stirring a breeze with her peacock fan and talking.

"I'm ver' bad womans, Madame," she'd say. "Not gotting any babies."

"Babies?" I'd trail along, half-asleep.

A long silence, then feigned laughter. "Babies like bugs . . . always making noise."

I knew what she really wanted to say, for she loved children, and when her hand clasped mine there was no East nor West—just Woman.

"Oh, you'll have babies some day, Mari," I'd assure her.

"No Madame, I'm old womans now."

She was just twenty.

Dos, I think, felt his marital status keenly. When safely out of earshot of his wife he was fond of remarking that for twenty-four summers life had smiled on him, but that during the last two it had laughed out loud *because he'd been married to Mari*. This was just a pose, of course. Some months later when the crisis came, their love showed fine and strong.

Dos was cook, his chief worry being to keep the larder stocked with chickens and eggs, our only fresh food aside from fruit. The village contained plenty of both. Nobody would touch them, however, as it was considered irreligious to kill meat for eating, and, to quote Mari, "They keeping eggs for making childrens, not selling."

The answer was periodic foraging parties along the highways and byways. When I spotted something edible in the chicken line, Dos ran it down, I gave someone eight annas (sixteen cents)—and that's all there was to it.

My maid and I were the natives' delight. As I wore trousers like the Sahib and had long hair like their own men, they couldn't decide whether I was man or woman. Eventually, by fair means and foul, they discovered my true identity and from then on I was the eighth wonder of the Burmese world. Every time I ventured forth from the cottage a crowd followed at my heels. The parade commonly began with a few naked tots and barking dogs, quickly growing into a motley assemblage of wild men and boys, plus some young girls—the same who, a few days before, thinking me a nice young man, had flirted with me.

I had been in a similar quandary about the natives when we first arrived in Burma. In the larger centers both men and women dressed alike in brilliant sarongs (*lungyis*) and short white jackets. It was very difficult to tell them apart until I learned the simple distinguishing feature—a scarf bound round the heads of the men, flowers and ornamental combs on the women. The British Tommies puzzled themselves with the same situation when they took over the country in the eighties and found, to their amazement, that many of their prisoners were young ladies.

In jungle towns, however, where the natives wore their casuals, confusing the sexes was impossible. Here the women discarded jackets leaving nothing to the imagination, while a man's wardrobe boasted of no more than a long *dha* knife, a loin-cloth, and a pair of "permanent" trunks tattooed on his thighs. But whether in town or country, and regardless of clothes or their absence, it is the woman who wears the "pants" in Burma—as I was to learn on better acquaintance with these people.

## 25
### Meet Bimbo Brown!

Gyat was distinguished for two things—an ancient Singer sewing machine and a brand new baby elephant. Both kept the village literally "in stitches."

The little chap had been orphaned when his mother was accidentally killed by trappers. This left him pretty much on his own, which for Gyat meant "on the town." For many days he wandered ad lib about the community begging handouts, walking unannounced into peoples' houses, poking his trunk into everybody's business and getting into all kinds of mischief.

Nobody claimed him. He was public property. Perhaps the villagers considered him above private ownership, or maybe they were just plain smart. In any case, he owned the village; also the heart of every last person he took a fancy to—including me.

It was love at first sight for both of us. Whether he sensed my maternal instinct or simply liked the way I scratched behind his ears, he immediately adopted me as foster mother and became one of the family, henceforth known as Bimbo Brown— "the baby."

No longer did he mope about like a lost waif. He had "folks" now. He carried his ears a bit higher—perky-like—and his eyes were full of fun. I fastened a gay ribbon and a bell around his neck. He was mine! And from then on he stuck closer than a poor relation.

Hardly larger than a St. Bernard dog, Bimbo furnished more laughs than a barrel of monkeys. He was constantly at odds with his feet or his trunk which were always getting in the way. This may have accounted for the comic, half-flabbergasted expression

145

on his face. His trunk never tired investigating this bright new world, and when he wasn't turning over somebody's ox-cart, rice barrel or water chatti, he was upsetting himself—which pleased him no end. Then he'd come running home to Mama, bell jingling, and tiny pink mouth—just big enough for peanuts—turned up at the corners as if grinning.

Since play and eats were the only things he thought about, I kept a jar of cooked rice at one end of our balcony for him. Several times a day one would hear him shuffling up the bamboo ramp for a tidbit. Eventually the wear and tear proved too much for the flimsy affair. After that he had to *wish* the rice down.

Feeding became something of a problem after a time. Not that we didn't have enough food—Dos gathered fresh bamboo shoots for him twice a week—but he grew hungry at the most ungodly hours, having developed a habit of snacking about three o'clock in the morning.

The little scalawag didn't mind waking us out of a sound sleep, either. Thank God he hadn't learned to trumpet yet, though his shrill nasal squeal was sufficient to wake the dead as it was— invariably setting off one of those impromptu husband-and-wife debates so common in the wee small hours the world over.

"Barnum—are you awake?" I'd ask, by way of opening negotiations. "It's your turn to walk the elephant."

A muffled snort from the depths of his pillow. Blissful snores.

Another piercing "sque-e-e-el" from below, this one producing a violent eruption of hubby's bedclothes, followed by a loud demanding voice, "*Now* what's the matter?"

"Our child is crying for his rice."

"No child of mine—the beast! I'll give him what-for—" Barnum would make a noise as if getting up, and I'd begin to have hopes. But, he'd only be turning over and adding with a heavy sigh, "—in the morning."

Which hubbub had the desired effect below where we could hear the servants moving about in self-defense.

"How's the baby, Mari?"

"Him throwing trunk around like crazy, Madams," her reply would come. "Me fixing him rice."

Whereupon the proud parents, duty done, would sink back to sleep—the deep, deep sleep that comes to those who have an infant elephant on their hands.

# 26
## Jungle Days

Where "the dawn comes up like thunder," as Kipling wrote of Burma, one is hardly disposed to oversleep. Nor is the odor of fresh baked bread wafting up from the kitchen likely to keep one in bed. Between the two, we were up before five and rarin' to go. While my man slipped into his clothes I popped an egg or two over the charcoal fire, set the tea a-brewin' and rustled up some Grape Nuts. Maybe we'd have a big juicy papaya, too.

Breakfast was a private affair and over all too soon. A hurried peck at wifey's cheek. A whispered, "I love you," in return. Barnum mounted his old gray mare, pointed her nose in the general direction of the Pondaung Hills and was off like St. George for his daily tilt with some prehistoric dragon—if only he could find one. Thus far, the Burmese variety had all proved of the "reluctant" type.

I had hunting of my own to do. Husbandless and adventurous, I would sally forth into the village accompanied by Mari, my girl-scout, who always scouted with a stout stick, just in case. Not far behind would be my "shadow," wee Bimbo, half-asleep and wishing he were back in bed. He'd act lively enough, though, when it came to a romp with a pig or a go with an old she-goat. And did he love to break the family circle when it came milking time for the kids.

At this point the village slowly awakens, small sounds rising against the silence; a crowing rooster, a barking dog, some happy wife singing over her morning chores. The click of a hand loom tells me that someone is weaving a new skirt; the thump-thump of

148

a wooden pounder that someone else is going to have rice for dinner.

Young unsmiling monks in bright yellow robes file through the dawn, collecting their daily ration of rice from the villagers. Silently they move from door to door, heads bowed, black lacquer bowls extended. Never a word do they exchange with the giver whose only thanks is that which wells from within the heart. Then—like the mists—they are gone; back to the monastery, to their meditations, their teaching—to the Buddha.

The mangoes are budding, shedding their fragrance on the morning air. Plantains and tamarinds shade the dwellings—bamboo-thatched and high off the ground to cheat the termites. None have more than one story; 'twould be beneath a Burman's dignity to have someone living over his head.

Against the huts lean long, bamboo "swatters." You'd never guess their function. These are the village *fire-extinguishers*—not always too effective, for while they beat out the flames in one spot they fan them in another. Fortunately, the Burman is not given to collecting material effects. All of his personal fortune is either loaded on the good wife's person or is in the family chest, since both depositories can be thrown out the window at a moment's notice. Should the house happen to burn—why worry? His friends merely sharpen their *dhas*, make for the nearest bamboo brake and, amid laughter and general hilarity, hack out the wherewithal for a new home.

Bimbo liked to dawdle. This wasn't good because one never knew what the "orphant" was up to. Chances were that he was smeared to the eyes in toddy syrup, having just toppled over two jugs of same—the property of Maung Than (Mr. Million), the *toddy-wallah*, who was up a tree and could do practically nothing about it.

Mr. Than owned a grove of toddy palms on the fringe of town where he performed each morning in the treetops, gathering the fresh sap in jars slung from his waist.

When first drawn from the tree, the juice is a mild insipid beverage, but it rapidly ferments into a sickening firewater enjoyed by the Burmese on special occasions. Most of the sap is carried to a long palm-leaf hut and boiled down into delicious brown sugar. Holding a monopoly on sweets and drinks, the *toddy-wallah* was an important man about town.

At about this time people were beginning to appear on Gyat's main lane. Already some youngsters had started a ball game. *Chin Lon*, it was called—played with a wicker ball kept in the air with quick, skillful movements of every part of the body but the hands. They were just going strong when who should come barging along but Mr. Thadu with his brace of water buffalo led by his little son, age four. Did this stump the ball players? Not at all! The game went merrily on, but I beat a hasty retreat to the sidelines. Though a native child can lead these animals by the nose, the smell of a white makes a water buffalo see bloody red.

Mr. Thadu was a farmer. He worked a patch of paddy thereabouts. You could see him any morning in the gray dawn, wading knee-deep in his rice fields, cultivating the tender shoots. He and his kind are the backbone of Burma, where an entire way of life centers around this tiny kernel and its growth. On the side, Mr. Thadu dabbled in corn and tobacco.

Another important citizen was the proprietor of the oil mill. I often strolled down to his place to pass the time of day. Then too, it was fun watching the old blind-folded bullock that provided the power. Red bandanna tied coyly over its eyes, round and round the beast plodded to the creak and groan of the wheels as they squeezed oil from tilseed and peanuts for lighting and cooking purposes.

The customary buzz greeted our approach to the town well, the community center. This was woman's world where she daily came to discuss who did what to whom, and who paid. A threatened divorce had tongues wagging today. Wide-eyed matrons listened with gaping mouths to the juicy morsels offered by the village snoops.

Mari translated, as best she could, the rapid-fire innuendoes. Mr. X was leaving Madame X. Why? Because she didn't love him anymore! The case had been placed before the elders.

Yet Mr. X was getting cold feet and everyone knew the divorce would never come off. Ma had the money and, as the reluctant partner, she would keep the children, as well as all property accumulated during the marriage. Which left the poor husband holding nothing but his old kit bag containing exactly one frayed loin-cloth, a well-worn back-scratcher, and the pink *baung-gaung* (headscarf)—his original contribution to the marital partnership.

Here I learned that a wife could obtain a divorce on a variety of grounds including non-support, chronic illness of the husband or senility; that a man might be freed if his wife bore him no male children or frequented places of risqué flavor; and that strict property laws safeguarded capricious separation.

The sing-song of voices coming up the road prevented my gathering further details. Suddenly deserting their gossip headquarters, the women flocked toward an approaching group of natives jogging along under heavy loads and heading for the marketplace.

"Hurry, Madame," Mari called excitedly. "A travelling store!"

27

JUST AMONG US GIRLS

Drawing near were three stocky merchants, and one tall slim one with head-gear of fancy baskets— Bless my soul, a woman! Despite the many miles trudged since before sun-up, they seemed none the worse for wear—laughing and chatting as they unshouldered their burdens.

Wares unpacked, the travelling store was open for business. Pagan lacquers, Mandalay silks, parasols and sandals were priced and haggled over, and the jungle ladies, though infrequent shoppers, proved sharp bargainers.

However, the most coveted commodity of all carried no charge. This consisted of news and choice bits of gossip from the bustling cities and river ports. Being wise merchants, they parceled out their chitchat with their sales—the bigger the sale, the bigger the talk. Various items went something like this: "Holiday in Mandalay. Big *pongyi* funeral with monks riding in autos. Everybody shooting off face and fireworks— Murder in Kyadaw! Headman gets head cut off at dinner—loses appetite." Which last seemed to indicate that being an honest judge didn't pay—even in Burma.

Like travelling salesmen everywhere, they had their little jokes and snappy stories that set the women tittering, and primed them for further purchases.

"What are they saying?" I asked. "It must be good."

Mari covered her mouth with her hand. "No Madame—bad, bad, bad!" she said, thus bearing out the Burmese reputation for *double-entendre*.

152

Such is the bill these roving merchants fill in the small villages remote from rail and river. Not only are they the common-carriers, but also the newspapers, message bearers, lonely heart columns and want ads—the main stem of the jungle grapevine. Tomorrow they would have a new headline: "Americans in Gyat!"

The travelling store fell right into my lap. Here was my chance to repay all our friends for their kindnesses. "Mari," I said, "we're going to buy them out!"

The girl looked baffled. "But what for?"

"I've decided on a village get-together this afternoon at our *zihat*, and everyone present will receive a gift. I'll name the article and you do the dickering. See that lacquer tray with the six little bowls? That's for the *tedje's* wife. And those painted lacquer betel boxes—don't you like those? Let's get the wooden one, the one of bamboo wicker—and this precious number made of lacquered horsehair. That's for *you*, Mari." Grateful squeals! I continued, "For the kiddies, those toy animals and penny China dolls. And I think the Mammas would just love these wooden sandals with the flowers painted on them.

"Now for the sweet young things—you know what they like, Mari. Pretty gewgaws for the hair; the white bone hairpins and combs. Maybe some of those lockets, too, made of coppers set in filigree work. How about these safety pins strung on cords? What in the world are they for?"

"Using for charms," my servant explained. What kind of charms, heaven only knew. Something to do with pinning down a man, perhaps.

"Smokes for the men," I went on, "cheroots—the big brown kind with the tobacco. And don't forget the betel-chew, Mari; better buy all of it." No party would be a success without betel-chew,—the customary smear of slaked lime, cracked betel nuts, cardamons, cloves and other spices wrapped in a leaf of betel vine. I knew they'd mix their own, as usual, and stay happy so long as there was a "quid" in the house.

Our shack was fairly bulging with presents that afternoon as we prepared open house for the guests. It was also bursting with song.

"Oh, the things that you learn from the Yellow and Brown will 'elp you a lot with the White," I chanted, slightly off-key, while my maid hummed a loony tune from far-off Hindustan. No one knew it, but we had smuggled home a wee snifter of toddy; that is, no one but Bimbo. And he was layin' low and sayin' nothin'—I hoped.

"Better keep an eye on me during the party," I cautioned. "I don't want to mix things up, hanging g-strings on the girlies and safety pins on the papas."

So, with a drap o' moonshine, Santa in every corner, flowers all over the place, and both Mari and I blooming with jasmine, the stage was set for a wonderful time.

Our guests arrived, a bit previous, bearing gifts of their own—bunches of green plantains; papayas. They brought trays of Burmese dainties too ranging from wingless crickets, crisp and brown, to honey bees stewed in oil. There was curry of tamarind and hot-damn chilli, with quantities of pickled tea.

The jungle women and I got along famously, everything so open and above board, with the naked truth on all sides. The young misses came dressed to kill in bright red *lungyis* and crisp white jackets fastened with buttons of gay-colored stones—all modestly covered from head to toe. Not so their married sisters who had long since shed the jacket for baby's sake. In fact, everything was for baby, including rummaging through my personal effects for silk undies which seemed to fit, carbolic soap, and tooth-paste. This last they smeared on the tiny heads because it smelled so nice.

The Mammas were perpetual merry-go-rounds for the darlings and mother's milk, like everything in Burma, was being shared with the needy. Ma Thin did double-duty, one fat little dumpling tuning in, the other dialing for more. When not nippling, each dumpling was mumming a pacifier—in the form of a *cheroot*.

Sad to say, the infant mortality is high in Burma—about thirty per cent—because ancient custom and superstition supplant hygiene and sanitation. Anyone handy with herbs who thinks he has perfected a cure passes for a doctor, and old women of the village serve as midwives. As soon as a baby is born it is bathed in cold water, while the mother is steamed with hot bricks to drive out the evil spirits that have taken possession of her during labor.

As Mari summed up the situation, "Burmese womans ver' good—working hard, making plenty moneys. Burmese mans no good—only laughing and making babies."

At least there's nothing wrong with this last recipe. Burmese babies are the fattest and best-natured in the world. The only thing that ever made one of them cry was the sight of me.

When my husband came home at sun-down, the party was still in full-swing. Our greeting kiss drew roars of laughter from everyone, osculation being one of the few pleasures the Burmese have somehow overlooked.

With the last glimmer of light, the "tuc-too tuc-too" of a large tree lizard sounded taps, and Barnum answered the call much to the amusement of the natives. Although it was high time for them to leave, no one made a move in that direction.

"Maybe they'll take the hint if we go to bed," I whispered encouragingly to Barnum, knowing how tired he was.

After we'd snuffed out the last candle they just laughed and shifted feet, then sneaked up closer to peek through the cracks.

"Dos," I called, firmly, "tell our guests the party's over."

I could still hear them giggling expectantly as we fell asleep.

## 28
## "Coming-out" in Burma

The Burmese like nothing better than throwing parties themselves. Barely a week had passed since the Feast of the Rain Nat
when the villagers were preparing for another Bust. This was to be
a *pwe*, the popular form of entertainment that marks every event
in life. It is a combination vaudeville and minstrel, with songs,
dances, monologues and chorus all free.

One man foots the bill for the whole show. It matters not if he
goes into debt, for Burma is an easy land on debtors. The world
and his wife are invited and no one "regrets." All guests come prepared to eat, sleep and be merry for the duration—usually five days
and nights.

This *pwe* was to be a teen-age party to celebrate the ear-boring
ceremony which takes place when a girl becomes of marriageable
age, or about thirteen. It is the greatest event in a Burmese girl's
life, and they make it a gala affair.

I was wondering how I might take part in the doings when a
delegation of women, headed by the *tedje's* wife, came strolling up
all togged out in their Sunday best—a dazzling array of color topped
with flowered parasols "Would Memsahib please to come ear-
boring festival?" the wife of the headman asked.

Would I! But what to wear? I had an inspiration. What could
be more fitting than the Burmese costume I bought in Rangoon—
plus, of course, a few native beauty aids to add the proper touch of
glamor?

That was where the fun began.

156

I reached the jungle beauty parlor, a palm shack balanced fifteen feet in the air on four stilt legs, by means of a jittery bamboo ladder, the sole means of access. Teetering amidships, the thought flashed on me that Darwin was right. I kicked off my wooden sandals, and during the monkeyshines that followed I needed only a tail to convince anyone of evolution. Frantically clutching my clothes, I finally made the top on all fours—a reasonable facsimile of an upside-down cake.

A giggly Miss showed me to a matting rug on which I reclined while the cosmetic artists did their darndest to make a Burmese beauty out of an American Plain-Jane.

Dried bark, ochre-colored, ground on a flat stone and mixed with water and a dash of scented talc—presto! face-powder. A smear of henna for ear lobes, finger and toe nails; crimson stain for lips—all applied with a finger technique. For the final touch my tresses, lustrous with coconut oil, were swirled in an up-swing hair-do and adorned with jasmine stars.

"*Hla, hla,*" ("pretty, pretty,") they cried as I regarded myself somewhat doubtfully in the shiny end of a kerosene tin.

"You sure have gone native," I gasped at my reflection. The effect was that of a fried egg with a busted yolk!

While all this was doing, I listened to some valuable tips on the Burmese idea of beauty—which is slimness at any price. And what a price they pay! For example, a young girl just budding into womanhood has hot bricks applied to her breasts to retard their development.

The average jungle Miss needs little to enhance her natural charm, and, for the most part, displays a refinement and simplicity of taste altogether refreshing in the over-dressed Orient. With small supple figures tapering to dainty hands and feet, with hips sinuous even by American standards, theirs is a decidedly streamlined conception of the body beautiful. The curve of their bodies framed in a bamboo doorway is easy on the eyes, to say the least.

A loud fanfare announced the commencement of ceremonies as I took my place beside the parents of the girl. This was a great day for them, as it marked their daughter's coming of age.

While she sat beneath a plantain tree, probably wishing she were a boy, the girl's ears were pierced with long gold needles, her outcries the subject of much to-do among the oldsters. Her cries would determine whether she had the makings of a good wife. Lusty wails received jubilant applause; half-hearted whimperings frowns and wagging of heads.

The serious business of ear-boring over, everyone made merry. Amid general rejoicing, the young "deb" took her bow before society, emerging from the bright cocoon of childhood into the golden springtime of her years. Our hosts served drinks—coconut milk and toddy in carved silver bowls—and a Burmese band gave out with jungle jive.

In the center of the *saing-waing*, a small circular enclosure resembling a baby's play-pen, one frenzied musician with turban askew and fingers flying banged out his special brand of noise over a series of graduated drums. His would-be accompanist, in a smaller *kyi-waing*, saved wear and tear on the fingers by using a wooden mallet which he pounded with ear-splitting delight on many-toned brass gongs.

Blaring trumpets and braying flutes added their dash of din to the musical potpourri. A tom-tom boomed—cymbals clashed—clacking bamboo clappers mingled with the notes of a primitive bamboo marimba.

Came the vaudeville sketches with jugglers and comedians; also dancing girls who curiously didn't dance with their feet, which seemed glued to the ground, but with the upper parts of their bodies only. The ballet moved as one to the rhythm of the orchestra—posing, posturing, swinging and swaying with rippling undulations of arms and shoulders.

This continued so long that at last I felt something must give—and hoped it wouldn't be me. Then, without warning, they broke into wild abdominal abandon, tempo hoochi-cooch, and leaped into the air as if stung by a bee. The music shot off-key in a final crescendo—and suddenly they were back on the glue again.

Inevitably, both audience and entertainers wore themselves into a state of happy, hungry exhaustion. Whereupon they unrolled matting rugs, and the tired celebrators awaited the long-anticipated feast.

Its main fare consisted of rice prepared in many tempting ways, most of it mixed with pork in a curry, or made into rice puffs and rolled in syrup.

The *pièce de résistance* was something which the Burman always reserves a place for at his banquets—that doubtful delicacy known as *Ngapi* (the *g* remains silent, but it smells as loud). Rumor had it that this item once was fish—preferably cat or dog fish which has no scales—but at the time of eating it resembled nothing that was ever alive. After being caught, said fish had been left to attain a *ripe* old age in the sun, becoming, in the process of decay, the celebrated fish-paste of the above name. For my part, the sun was its undoing.

At last festivities were over, with the little lady undoubtedly feeling several years older. Well she might, for with this celebration she was now ready for the affairs of the heart—and the established "courting hour" between nine and ten each night of the full moon.

This is youth's sweet hour of romance when Burmese boy meets Burmese girl, and dreams are made that only wait on love for their fulfilment. Someday her prince would happen by, and when he did she would not be denied. Love is a wild impulsive thing in the tropics—impatient of compromise, of delay, of tomorrow, of everything but complete and unhesitating surrender.

Should the boy of her choosing prove indifferent, the thwarted maiden may resort to any one of several remedies guaranteed to rouse the faint of heart. First, there is the "Love Potion" of herbs. If this fails, then she herself may, in desperation, take the "Death Potion." Happily, the beloved usually weakens and is presented forthwith with the "Fingernail Charm," a long gilded fingernail grown by the girl for this purpose, which he wears forever about his neck—except when using it for a back-scratcher.

If parents approve the match, a *pwe* is held for the united couple, and they begin life as man and wife. In case parental consent is withheld, the lovers have no choice but to elope. As Mari explained, "Not making any marriage, only running away together."

But it's no mere escapade. This act is binding in the eyes of the community. Upon their return from the usual three or four day "honeymoon" in the woods or neighboring village, the lovers are considered duly wed. Just an old Burmese custom!

In any event, there is no marriage ceremony, as we know it. According to the Buddhist faith, all things are transient. Hence we cannot lay up treasures on this earth, even in the shape of husbands. Nor are marriages made in heaven. They are of the earth—earthy, and the Church cannot bless any earthly ties.

Nor does the State play any part in the union. Marriage is purely the personal concern of two people sharing a mutual desire to live together. Love is the only bond.

Wedlock is a fifty-fifty proposition from the very start, the bride retaining her individuality, her property and even her name. Should you ask whose house this is in Burma, they will tell you it is the home of Miss Jones and Mr. Smith—though man and wife.

After the first flush of connubial bliss the newlyweds settle down and the wife launches herself in business. Practically all of the country's petty trade is in feminine hands, and rare is the woman who cannot boast a tidy profit from a stall in the local marketplace or a shop in her own home. Whatever vocation she follows, however, the Burmette is first of all a homemaker and mother.

Life is not all work for the housewife, though. When the husband is home, alternately doting over his young 'uns and pottering about his paddy field, she manages a few moments of leisure.

Her chief form of amusement, and the indoor sport of most Burmese wives, is dressing the husband. She takes great pride in having him the cock of the walk, and devotes entire evenings to fussing over his clothes, arraying him in the finest silks for his nightly rounds with the boys.

Actual divorce is virtually unknown. Marriage, founded on love and sustained by mutual faith and understanding, is a most successful institution in Burma. "For better or for worse," invariably it lasts for the Burmette and her man "till death do us part." Together they achieve an ideal way of life that makes Burma little short of paradise. Living, they believe, is not a business. Living is an art—and certainly a pleasure.

## 29

### Tattoos and Toddy

Just as Burmese girls have their ears pierced to signify their coming of age, so the boys take it out in tattoos—the marks of manhood.

All conscientious parents must see that their male offspring acquire that well-scribbled look before the middle teens.

Junior is taken to the local *saya* (tattoo artist) at an early age where the point of a two-foot needle quickly acquaints him with the first step in the painful process of growing up. It's a pretty terrifying business for the little fellow at this stage; something like our first visit to the dentist. But it's the custom and Mom knows best. Furthermore, if he doesn't stand still and behave while the nice man is poking his rear, he'll be walloped—or worse, the bad nats will get him. If the kid doesn't respond to the lecturing, he does to the opium which is administered forthwith internally and proves an effective antidote for stubbornness.

However, since this initial design is applied to the fannie, a bit of child psychology usually precludes the need for opiates. The parents cook up a neatly contrived cock-and-bull story about the tattoo being a sort of vaccination against spankings and other such dread afflictions caught from older people. The gullible tike swallows it whole, sweats out the ordeal and everybody's happy—until the next paddling when he discovers that he has been grossly deceived.

With this initiation into the male fraternity little boy becomes big boy, and his visits to the *saya* increase. The maestro goes to

161

work on his thighs now, lacerating the flesh and filling the cuts with dark blue and red stain made from plant sap. Dispensing now with childish fictions, he informs the youngster that it is his "fortune" which is currently being scratched into him.

Henceforth, he will have at his fingertips a ready-reference guide to the future, with special charms to offset evils. No longer will it be necessary to consult the village astrologer for his lucky or unlucky days; a peek under his *lungyi* will do the trick.

Of course, it takes an acrobat to read some of the stuff. But, when a man gets around to that part of the design he frequently has a wife to do his research. And sometimes there's astrological data between the lines that makes most revealing reading. Any married Burmese woman will vouch for this. No secrets here. If she's ever in doubt about her husband's past or future, all she has to do is check on his thighs.

This tattoo business continues for several years, during which time the proud victim blossoms out in a wonderful rash of fancy animals, scrolls, doodlings, glyphs and maybe parts of the Morse Code, etched all the way from waist to knee. There isn't much space left. The fellow is equipped to give a Broadway leg show. And one day the maestro gets around to drawing in his client's obituary, so it's all over.

The Burmese lad can now assure himself, following a serious study of the asterisk on the left knee, that manhood approacheth. So the man-to-be sallies forth into life, exposing his bright new zodiacal girdle to the girls at the least provocation—all the while with his eye on that cute little wife and tobacco farm clearly inscribed just below the navel.

But youth is impatient. Although each leg resembles a damascened gun barrel and his midriff presents a veritable cyclorama of things to come, true manhood is not yet. For manhood is not merely skin-deep in Burma; it is also a matter of the soul. He is, as yet, but half a man—the physical half. Fifteen years he has given to the development of the body. Now the time has come to cultivate his spirit. The Burmese youth faces the final barrier into Man's World—his novitiate in the monastery.

Two such candidates for manhood passed into this phase of their lives while we were in Gyat. Their entry into the monastery occurred during the town's most festive holiday, the pagoda feast.

A pagoda feast is a glorified church picnic, only more so. There must have been a thousand people gathered in Gyat for the occasion, and the tiny village was bursting at the seams. They came winding over the hills from miles around, bearing gifts for the monks and contributions to the party. The devout arrived for religious reasons; others to see the novices; some for the free "eats"; all to visit with old friends and be merry.

It was full-moon time, and the boys and girls took full advantage of it. Dark glades echoed to familiar squeals and the beat of myriad feet. In the village one heard the thump of rice pounders mingling with the rich laughter of married folk.

All through the night it went on—people coming—people going—lugging baskets of green tomatoes, papayas, plantains and great fat pigs for the feast—crowding around the sputtering fires. They sniffed hungrily the huge sizzling kettles that filled the air with odors of pork curry, and sweet rolls made of rice flour squeezed into boiling oil through a hole in a coconut shell.

All of which made for some delectable eating the following day. And just about the time when everyone was foundering and looking for a place to nap the ordination ceremony got under way.

Somebody started in on the gongs again, and a long procession commenced its slow solemn march through town. In the lead walked the two little novices dressed in gay attire, their faces covered with ashes to denote the passing of earthly joys. On either side of them a large *hti* (parasol)—one open, the other closed—represented the fullness of life they were enjoying and the shutting off of worldly pleasures in the life they were about to enter. Over his shoulder each carried an umbrella and a pair of sandals—the sum total of his material possessions. Gift-bearers followed, some twenty women with baskets of peacock feathers, flowers, fruits, saffron robes and rice, not to mention ample sweetmeats for the monks.

The party halted under a large palm canopy shading a raised platform at the far end of which stood four huge palm-leaf fans. One by one the women stepped forward to place their offerings on the stage. This was the signal for lowering the fans, behind each of which was a monk, his meditations completed and apparently feeling no further need for keeping out distraction.

To the sing-song drone of prayers, the two new souls were given to God. A skein of white cotton, placed over their heads, was then wound around the Holy Men, thus binding the youths to the church.

They spent the rest of the day with family and friends. At eventide they said goodbye to all they had known, and received the yellow robes of priesthood. In the morning they would open their eyes on a new life of study and spiritual awakening.

For two years these young boys would live away from the world, deep in their own thoughts and the ageless wisdom of the East, learning the tenets of Buddhism. At the end of that time, being spiritually grown, they would be given the choice either of returning to society as full-fledged men or remaining in the order—eventually to become monks.

# 30
## In a Buddhist Monastery

As with most Burmese monasteries, the one in Gyat stood off by itself a short distance from the village—a large, oblong, single-story affair of brown teak sprawled in the shade of innumerable trees.

Women are seldom permitted in these holy-of-holies, and it was only by the grace of God and the persistence of the *tedje* that the head-*pongyi* (head-monk) granted me a dispensation.

Allowing time for the novices and lesser clerics to scurry out of temptation's way, I arrived in the afternoon, first removing my shoes, as is the custom, before mounting the steps to the verandah. Two hideous creatures, half-man, half-lion—and fortunately plaster—barred the way for demons and evil spirits. I passed, however. At the head of the stairs the *pongyi* waited, a young-looking man with shaven head, wearing a saffron robe wrapped about him like a Roman toga. He beckoned me inside.

In sharp contrast to the ornate exterior, with its multi-tiered roof, carved-wood eaves, ornamented peaks, spires and frilly dodads stuck all over, the interior was simplicity itself.

From the doorway a long hail led back—bare and gloomy. Apparently, it served as a dormitory at night, for along either side, in neat rolls, the monks had arranged their matting beds. Directly opposite, against the rear wall, near an alabaster image of the Buddha, set two straight-backed chairs with a full-length mirror between them. Some old chests stuffed with manuscripts lay open at one side, and there were wooden statuettes, also copies of the *Times*

165

*of India* and the *London Illustrated News*. The walls themselves were covered with paintings ranging in subject from the coronation of Queen Victoria to a group of yellow-clad monks meditating in some pea- green woods. A portrait of the Holy Family hung beside one of the elephant-god, Ganesh. In fact, every religion was represented, all jumbled together in most friendly fashion.

The monastery kitchen looked trim as a new pin, its plank floor scrubbed smooth and bright. Long low tables held stacks of pink and blue china bowls, trays of lacquer and brass, rice jars, knives and spoons. Occupying one corner was a huge barrel hollowed out of a teak log. This contained left-over rice for the poor. Next to it I noticed an open brazier where the food received daily from the villagers was heated for the monks.

They didn't permit me—a mere woman—to investigate beyond the kitchen. The rest of the monastery was forbidden territory. Smiling, the *pongyi* guided me back to the central hall, where we sat and talked quietly.

Once, during a lull in the conversation, I heard voices coming from another quarter of the building—children yelling out their lessons at the top of their lungs. There was a reason.

"That will be our pupils from the village," the friar explained. "We believe a youngster memorizes quicker by ear—hence the noise. Our curriculum—?" He let his tongue rest on the word, then rolled it over in his mouth as if savoring the taste of its syllables. "Our curriculum," he repeated, "includes history, geography and arithmetic.

"Of course, the *novices* engage in *higher* studies. At present, they are learning the duties of a monk. Their day is divided into three study periods—first after the early morning meal; second following the regular breakfast gathered in the village, and once again after the last meal at noon-time.

"Oh, yes," he chuckled, "anyone going through his novitiate must live on strict schedule—up at sunrise, then to his chores of sweeping out the monastery or washing down the floors, carrying in fresh drinking water and tending the sacred trees. At mealtime he waits on table, cleaning up the place again in the afternoon. By the time he reaches monkhood, good habits are a part of him. His life then becomes one of teaching, meditation, reading the sacred books, and—when he is old—resting.

"Usually," he rambled on, "our day is fairly well finished by four o'clock when classes are dismissed and the children sent home. After that, the monks and novices have till sunset to stroll about the village or countryside."

"Till sunset?" I put in.

"None are allowed out after dark," he explained. "In the evening, I gather the student-*pongyis* together and examine them on this or that. Sometimes I give them a little talk. At nine, we have what you would call 'vespers,' and then to bed."

"A busy day!"

"Too busy!" he agreed, quickly. "I am always glad when it is over, for only then do I find time for my own meditations. These are very necessary to a Buddhist monk, you understand. Meditation holds the most important place in our religion."

"And what do you meditate on?" I asked.

"I think back on the life of the Lord Buddha, following over its course again and again so that I may the better pattern my life on it." He paused and looked at me with a hesitant, "Do you know the story of the Buddha?"

Without waiting for my reply, he settled back and eagerly unfolded the high lights of a man born nearly twenty-five hundred years ago, whose teachings a third of the human race today regards as the highest truth.

"The Buddha," he said, "came into the world a Hindu prince, heir to a kingdom in Northern India. His name was Siddhartha Gautama. Though strong of body and quick of mind, it was early noted he seemed strangely discontented, caring nothing for the pomp and luxury of court. Even after marrying the beautiful Yasodhara whom he loved, a sense of futility continued to plague him.

"Then, one day in his twenty-ninth year, he came upon an old man, a sick man, and a dead man in quick succession. Heretofore shielded from such sights, the eyes of the pampered prince were opened to the misery of life and his heart filled with compassion.

"It was this sympathy for the sufferings of mankind that prompted him to give up his life of ease and go out into the world. It is said that he left on the very night of the birth of his son."

The monk let his eyes wander down the hail. "Think how it must have been for him that night. Remember, he was still a man—not the Buddha—still weak and human. We call it the night of the Great Renunciation, for it was there in the semi-darkness that he renounced everything to become a common man and himself know the misery of life so as to find its answer and perhaps its cure.

The *pongyi* brushed a delicate brown hand over his shaven head and resumed, "Years later while resting under a Bo tree one night, the truth flashed on him and he saw life and human suffering to be one and inseparable. He saw our present existence as but one of many cycles of pain through which man must pass to gain Nirvana— Heaven. With this, Siddhartha Gautama became him whom we know as 'The Enlightened One'— the Buddha."

"Then what?" I said, as he paused.

"Leaving the search, he then went forth to teach. He urged us to love one another and live in peace, and advised moderation in all things. He left five major commandments which everyone should obey. Let me repeat them: thou shalt not take any life whatsoever; nor steal; nor commit adultery; nor lie; nor imbibe intoxicating drinks. These are the principles, Memsahib, which today guide the lives of five hundred million people.

"For the monkhood," he added, "the Buddha laid down five additional commandments. We are forbidden cosmetics or personal adornment; singing, dancing or making music; eating after midday; overly comforting the body or accepting gold or silver."

A dull booming sounded somewhere in the monastery, echoing through the corridors and filling the central hall. The *pongyi* arose with alacrity. "That is the *kaladet*," he explained "—the wooden bell rung each day at dawn and sunset, like your Angelus."

He accompanied me to the porch and said goodbye.

# 31
## Granny Ti's Exit

That the time should ever come when Gyat wasn't fixing for a party or sleeping one off was inconceivable. Yet, that is what happened.

Whether due to some oversight on the part of the Horoscope Committee or because there weren't enough births, ear-borings, rains or droughts to piece out, a decided gap opened up in the entertainment schedule. The citizenry were anything but happy about it, the town council was at wit's end, and everybody was blaming the *tedje*. Woebegone and looking thinner, the little man came to us.

"Village having much need of *pwe*," he announced sadly, the catch of pleading in his voice.

Barnum stifled an exclamation. "But you just had one a few days ago."

"Two weeks passing since pagoda feast, Sahib," was the indignant retort. "Two weeks long time no having *pwe*. People thinking I no-good *tedje*."

My husband's arm encircled the chubby shoulders. "You're a first-rate *tedje*," he said. "Something will surely turn up before long." His eyes flashed with sudden inspiration. "What about those nats of yours? Couldn't they be made to help in a situation like this? I mean— couldn't you sort of get a message from them ordering the villagers to hold a *pwe* for the Sun Nat or something?"

The headman shook his head. "Me already asking nats. They not hearing . . . maybe not wanting *pwe*." The pleading crept back into his voice. "Only you and Memsahib can helping."

"We—help? But how?"

"Finding big bones—right-away, quick, now—so we can making celebration."

Barnum smiled sympathetically. "I wish it were that easy," he sighed. "Believe me, if we could, we'd get you the biggest batch of bones in all Burma—just so you could have fun. But . . ." and he went on to acquaint the disheartened native with the elusive nature of fossils.

That was how the matter stood—till the news came that saved the day, and possibly the headman's job. The *tedje* had all but given up hope, and things were going from bad to worse when—God bless her soul—Grandma Ti obligingly dropped dead. The problem was solved!

Now Granny Ti was old—some ninety years plus. This detail, however, hadn't kept her from trying to be the village belle at the last feast, and she died in the attempt. Two nips and a naughty did it. The "naughty" she might have survived, but the "nips" were 100-proof toddy.

The story was that she had always wanted a big-name band for her funeral and, since the "tuney crooners" were in town, this was her chance to do and die, which she did—to her everlasting glory. It was several years since anyone in Gyat had been fortunate enough to pass away with an orchestra on hand. Granny was the envy of the village.

The affair had Barnum and me flabbergasted, until we were made familiar with the facts of death in Burma.

It seems that dying in this strange land is one of the chief pleasures of living; people spend their entire lives looking forward to the end. Life is one vast show, full of laughs and jokes, with the grand finale capping the climax as, amid the applause and approbation of his fellows, the Burman gracefully makes his exit.

At the *tedje's* request, Barnum, Dos, Mari and I were invited to Granny's farewell party as guests of honor.

Night had fallen when we arrived at the scene of the wake where some two hundred souls were eating, drinking and making merry in the dim torch-light. For four days and nights they had been enjoying themselves at Granny's expense, and this was the last. Through the open doorway of the palm-fringed cottage passed a constant stream of guests laughing and chatting excitedly, all intent on one last glimpse of the departed.

I felt a slight uneasiness as we stood in the entrance, knowing that the highest tribute a Burman can pay a friend is a rousing good funeral. Inside, we wedged ourselves into the packed humanity milling about in a fog of burning wax, incense, cheroot smoke and sweat. In one corner, talking with a young yellow-robed monk, was the *tedje*.

We caught his eye. Beaming, he elbowed through the crowd toward us. "Big *pwe*. Big *pwe*," he chortled expansively, handed us each a huge cheroot and led the way to a small alcove where we were plied with the usual sweetmeats and drinks, including *le' pet*— a pickled tea made from tree leaves mixed with garlic, salt, oil and millet seeds.

More pushing and shoving as we worked our way to the front of the room where the throng was thickest. There, behind a filigree screen, in the flickering glow of candles, reposed the only silent member of the party, the little lady herself. She was laid out on the ground, all but her head buried in ashes—an effective means of embalming if the corpse doesn't dally too long. Clamped in her jaws was a silver coin, passage money, no doubt, into the afterlife.

Granny Ti had not been without funds when she died, the *tedje* explained. The very poor had to get by with a betel-nut between their teeth, or, at best, a lead coin.

For a time we stood there, listening to the headman as he described how the body had been prepared for burial—thoroughly bathed, the big toes fastened together with a strand of the eldest son's hair, then dressed in its Sunday best and placed in the ashes.

At one point I turned to Mari with the question, "Where do they think the soul goes after death?"

She replied in her graphic way, "If good mans, going upstairs with God—always eating and making happy. If bad mans, staying downstairs—all-time rounding, not stopping anyplace, not gotting any home, hiding in woods and making people fright."

One would hardly have recognized the orthodox Buddhist hereafter in this description, but it did point up the fact that a Burman
just can't lose. Even a bad one hasn't such a tough time on the
other side. That is the nature of these people—to be lenient with
their sinners and let bygones be bygones in death as in life.

Once during the evening, the young monk called for silence and
gave a brief talk. "This world is a bridge," he said. "Pass over it but
build no house upon it. Death is the entering into the Great Peace,
and when the hour has come for you—" he paused and swept his
listeners with a look "—remember your good deeds, for these will
be more to you than all else." A silence followed, when the room
seemed strangely empty, presently filling with the sounds of re-
newed revelry.

Outside, the big-time orchestra gave out with the music, just
as Granny would have wanted it. Tom-toms boomed, gongs banged,
and I could almost picture the blithe spirit of the little lady, freed
at last of the fetters of the flesh, cutting a final caper and waltzing
off along the Milky Way.

The following evening at sundown, they lowered a small red
and green coffin into a shallow grave in a lonely clearing on the
outskirts of town. They threw rice in after it; the warm earth covered it; plantains and water were left there—small somethings to
tide Granny Ti over the journey into the next world.

When we left, the cemetery seemed like a deserted fairground
from which the revelers had hurriedly departed. The place was lit-
tered with broken frames of fantastic shape, bones, faded strips of
cloth scattered in wild profusion. And from a mound of fresh earth,
one bright new wooden spire flaunted its gaudy streamers toward
the sunset.

We felt sure that all was as Granny Ti would have wished.

# 32
## Luminous Spider

One evening dinner was ready—but no sahib. Twilight blackened into night. Still no sahib.

"I'm worried," I confessed to the servants as we hurried to the headman's house. A thousand fears kept rushing through my mind. Barnum might be lost, thrown from his horse and hurt, set upon by some wild animal.

"Can't you send a search party to look for my husband?" I pleaded with the *tedje*.

He shook his head: "Burmans not going away from village at night."

Those nats again, I thought.

"Knowing which way Sahib go today?" he asked.

"Toward Thittadaung—quite far."

A frown crossed the pudgy face. "That bad place. No good village. Dacoits."

"Dacoits? You mean bandits?"

"Yes, Memsahib. Bad stealmans. Not killing people, only robbing."

That last offered small comfort.

At the mention of dacoits, Dos changed color and disappeared. Somehow, I didn't expect to see much of him for a while.

Further pow-wow with the headman disclosed the reason for Thittadaurig's evil reputation. It started back when the Burmese kings were making things—mostly necks—snappy, and dacoits were plentiful. One evening a lone white prospector pulled up at the

173

peaceful village. He was tired and jungly and, in no uncertain terms, demanded food and water. These not being to the queen's taste, he hove into the bunch and beat them up.

He was quietly sleeping in the still of night, when a few natives stole up and meted out justice from his belongings. Then they gave him the "bum's rush." Clad only in a coat of tan and a teak leaf, he fled the place, cursing it for a den of thieves.

His drawers, according to rumor, long waved on high above the chieftain's house in token of what, to the citizens of Thittadaung, had been an historic encounter between the Brown and the White.

My spirits slumped to absolute zero when I returned to the cottage. There Mari and I were left to wait through the night.

"Madame please not to worry," she consoled. "Dacoits not making troubles with mens anymore; just womans. Tying naked to trees and holding for ransom."

Somehow, this worried me even more. It was bad enough to have Barnum wandering through the jungles without his pants, but naked girls bound to trees along his path . . . that was a ticklish situation.

Shortly before midnight, four of the headman's deputies arrived with orders to guard us till morning. Their ideas of protection were unique.

I can only describe their antics as jungle boogie-woogie, though it could have been the Burmese version of the strip-tease, male style. No sooner had they entered the cottage than the fat one plopped himself in the center of the floor and commenced to shimmy and shake so hard I thought the place would fall apart. Off to one side his partner turned hand-springs. They both then got together in a war dance, screwing up their faces and singing at the top of their lungs about warriors and spirits of the forest. This petered out when number two boy made as if he were collapsing, and in place of the war dance we were treated to a solo.

The other pair, far more reserved, spent most of their time hugging each other in a corner, and grunting. My maid recognized this as native jiu-jitsu.

All four finally joined hands in a rousing ring-around-the-rosy, at regular intervals dropping to the floor, one by one, and bellowing blood-curdling threats through the cracks.

"What on earth are they doing?" I asked Mari. "Is someone below . . . or is it just their idea of having fun?"

"Oh, no. They scaring away devils," she squeaked from a neutral corner.

With all their fear and superstition, the natives were more frightened than we. This was their way of whistling in the dark. As the first streaks of dawn broke the horizon, home they went, completely done up.

My fears were now terrors, and I was well-nigh frantic when I heard the distant clump-clump of hoofbeats and saw coming through the morning mist a tired speck of humanity. The Sahib was home!

"You're sure the answer to a maiden's prayer," I wept, as Dos pulled off his boots and Mari poured a steaming cup of tea.

"Were you really afraid last night when I didn't show up?" he queried wearily but gently.

Mari cut in. "Oh, no. Madame not gotting any fright. Four men staying with Madame all night."

"Which brings up the other side of the question," I remarked, beginning to feel a bit maudlin. "With whom did you spend the night, Mr. Bones?"

Barnum fumbled for his pipe. "Well, I'll tell you," he began. "It was growing dark and I was heading for camp through the hills when suddenly, only a few feet away, I saw a ball of light big around as my thumb. Naturally, I thought it was a huge firefly—there were so many of them flitting about. But this light was different. It kept absolutely still. I dismounted and pressed the brush aside till I was directly over it. Then I struck a match. There in full view, Pixie, was a luminous spider, the large oval abdomen glowing in the dark. Still it didn't move. When my match died out the spider's glow increased again to full power.

"Fearing it might be poisonous, I wrapped a handkerchief around my hand and made a quick grab. The cloth caught on a twig and my treasure scurried away. I watched the light vanish in the brush and spent the rest of the night looking for it. Never saw the thing again." My husband sighed and looked up with an air of resignation.

"It would have made big news in the scientific world," he reflected. "No one has ever caught a luminous spider, and but one other white man, to my knowledge, has ever seen one."

## 33
### "GOD SAVE THE KING"

Gyat is the place where at long last I actually witnessed an example of human behaviorism that I'd read about all my life but never fully believed. It started with the appearance of a neatly-dressed little Burman whose dark skin shone with recent scrubbing.

Neither Barnum nor I had ever seen him before, and from his manner he obviously was not a product of the town.

Dos introduced him as Ba Tu, the bearer of an English gentleman lately arrived in the neighborhood. At this the fellow smilingly produced a card which bore the name of S. Alexander Packinham—Correspondent; and under it, the British news agency for which he worked. On the reverse side, in very precise penmanship, he had written:

> "My Dear Mr. and Mrs. Brown:
> May I have the pleasure of your company at dinner tomorrow night at eight? My camp is situated just back of the monastery in the toddy palm grove."

It took but a minute to scribble a reply saying that we'd be delighted. The small man accepted our message with a bow and sped back to his master.

Outside a large and expensive Muir tent, waiting to receive us the following evening stood S. Alexander Packinham—Correspondent. Was he dressed in the rough-and-readies of a Richard

Harding Davis, a Stanley, an Ernie Pyle? Did he look at all like a newspaperman? He did not.

We both stopped dead in our tracks when we saw him. My husband looked, blinked, rubbed his eyes and looked again. For a moment I, too, thought I was seeing things. But it was true. The man wore a white dinner jacket, black tie, and knife-creased trousers falling to glistening black shoes.

For an instant I wanted to drop through the ground. Barnum had on his usual field togs, while I featured checked gingham.

"What ho! there," the Englishman called out, striding up. "By jove, I'm glad you've come. Rather irregular of me, popping up like this and expecting you to come to dinner, not having met you and all that. But I wanted deucedly to see you two. And one can't stand on ceremony in these parts, eh what?"

We agreed, and felt somewhat better. He then ushered us into the large tent where the evening began over apéritifs.

Mr. Packinham had heard of our work and was all questions. Barnum's account of the India discoveries and those we presently hoped to make seemed to fascinate him.

In due time, the little Burman of the day before appeared from behind a flap, to announce dinner. Our host rose and showed us to a table set up in the center of the tent and spread with really spotless linen, including napkins of fine Irish cloth. The service was of silver, heavily monogrammed; the china, Wedgwood, arranged around a centerpiece of flowers.

Roast suckling pig formed the main course. As the tender, juicy slices of meat were carved and heaped upon our plates, Barnum remarked, apologetically, that had we known we should have come attired somewhat differently.

S. Alexander guffawed. "Jolly well glad you didn't," he replied. "I like my guests to be comfortable. You see, though it is quite an occasion having such a distinguished scientist and charming lady to dine with me, that isn't what prompts all this er—ostentation, as it were." He indicated the trim interior of the tent, the polished leather, bright metal; everything in place with military precision. "This is the way I live. All of the time, I mean."

His man set the heavy Wedgwood plates before us, steaming with richly-seasoned pork and fluffy rice.

He continued. "Yes, I dress for dinner every night, wherever I am, and Ba Tu trots out my silver and china. These are my home ties, and as I eat alone I think of—well, many things. The family place in Cornwall—the cottage in Devon—polo at Wimbledon—Imber Court and the trotting races—Piccadilly—my flat in Mayfair. All good solid stuff, you know—something a man can cling to in these deuced jungles.

"I suppose you might call it a kind of armor, this shell of home-stuff—tableware, manners, clothes, customs—with which we colonial British insulate ourselves. But it's the thing that keeps us going. Set that armor aside, and where are you? A man goes to pot quickly in the tropics, once he lets down.

"I've been knocking about the East over four years now, covering one assignment and another. This makes my seventh month in Burma, collecting material for a series of articles. When you've been away from home that long, living out of touch with civilization most of the time, you're apt to forget your standards and the things you normally live by. Well, this business of dressing up and carting around the family silver is my way of reminding me that I'm part of the British Commonwealth and way of life.

"Of course, there's something psychological about it," he went on, presently. "It's like keeping clean-shaven and polishing your brass in the Army. A man gets to feeling frightfully alone out here. A strange world for a white man. And unless he keeps remembering, he's a lost soul.

"Whenever I feel myself slipping, I say, 'Hold on there, old boy. You've got to go back to the British Isles someday, and you don't want to be a bloomin' barbarian when you do.' Then, maybe, I'll sing 'God Save the King' each night for a week, or take extra pains with something—just to stiffen up my morale.

"I know it's different with you Americans. You have the faculty of blending wherever you go. But we British must remain staunchly apart; otherwise we lose face—and control."

S. Alexander Packinham toyed with his water goblet, his shirt cuff snowy white against the deeply-tanned hand. "You see," he added earnestly, "we have never let ourselves forget the gulf between East and West. And, as we retain our original mode of life here because it's the only one for us, so we also permit the native peoples of the Commonwealth to retain theirs. That is why—though it may sound strange to you Americans—we have tried, in our colonial policy, to follow the maxim that the *least* governed is the *best* governed. We are here in the Orient to trade and develop resources that might otherwise lie idle. To do business, in short. Meddling in the private affairs—the religion, customs, and so on— of those with whom we deal, is strictly taboo." He glanced toward his servant who was standing by, silent and ready. "Ba Tu here is a good example. I would not want to change him. I like him the way he is."

The correspondent eyed us thoughtfully, then beckoned Ba Tu to clear away the table. The servant next placed lounge chairs outside the tent entrance, and to them we retired. The jungle edge loomed black, a tip of moon showing above the palm tops.

Our host snapped open a gold cigarette case and held it out: "Try a Piccadilly? Though really not as good as your American brands, they are hand-made. Part of my armor, you know." He winked at me and ho-ho'd.

"You were speaking of British policy in native matters," Barnum prompted.

"Righto." The Britisher eased back in his chair and breathed deep of the night air heavy with exotic perfumes. "Take the other day. I came upon a group of natives in a clearing on the outskirts of a village. They were crowded around an elephant—an old tusker lying on its side, thrashing about and bellowing in extreme pain. In its writhings a great pit was being hollowed out in the earth. The animal had been injured somehow and brought to this spot to die. The *mahout* (keeper) was there in the crowd, looking too sick to speak—as though he'd lost his best friend, you know."

Packinham pounded his knee for emphasis. "Not one of the blighters would raise a finger to put the poor thing out of its agony!

A slug from my .500 between the eyes would have done it neatly and painlessly. But I couldn't—no matter how my instincts demanded it. I couldn't have touched that elephant with intent to kill without violating the deepest feelings of those men. It would have been the same as killing one of their number who happened to be ill. As you know, they won't take life of any kind in this country; all the village butchers are Moslems from India, or Chinese."

My husband and I had noted the powerful express rifle hanging in its scabbard from the ridgepole inside the tent. "Do much hunting?" Barnum inquired.

"When I can. Brought down a big *Sambur* buck—the deer with the mane in the hills west of Mogok not long ago."

"Mogok?" Barnum repeated, sitting up. "Isn't that where the famous ruby mines are located?"

The other nodded. "Most celebrated in the world. That's where the priceless 'pigeon blood' stones come from. One Mogok ruby weighing 38 ½ carats is said to have brought 20,000 pounds sterling back in 1875. The Siam and Ceylon rubies can't compare with them. Thibaw, Burma's last king, sweetened his bank roll by fifteen thousand pounds from the workings. Good rubies are worth more than diamonds, you know. One of five carats might well bring three thousand pounds, while a diamond of that size would be worth only a tenth as much. They are much more rare than diamonds, too."

He flicked the ash from his cigarette, the tip glowing brightly. "As you have undoubtedly discovered, Mr. Brown, this is a wonderful country for minerals. There are scads of tin, jade and amber in the north, as well as one of the biggest silver-zinc-lead ore bodies on record. And, as for tungsten, there's more of that coming out of the Tavoy diggings south of Moulmein than I've seen anywhere.

"Deucedly interesting, too—that Tavoy district," he reflected. "Natives living along the coast down there keep domesticated pythons which they take fishing with them. These 25-foot reptiles have an uncanny weather sense, detecting the approach of a storm long in advance. When it feels a bit of a blow coming on, the snake plunges over the side of the boat into the ocean and heads for shore. You can bet the crew doesn't lose any time then putting in at some snug harbor."

The moon was high by the time Barnum and I made our way home that night. We said little. I didn't even expand on having at last witnessed the phenomenon of a Britisher dressing for dinner in the wilderness. But somehow the jungle seemed a little less wild, the world a bit better, and there was a fine, firm feeling inside of us.

I think it's always that way when you have passed an evening with a real *gentleman*.

## 34
### What's Best for Baby?

Globe-trotting has one drawback which sometimes outweighs all of its advantages. This was one of those times.

Barely had I begun to feel at home in Gyat and develop an affection for the natives, when up bobs my husband announcing plans for departure.

Departure so soon! It came with the suddenness of a lightning stroke, filling me with blank dismay. The thought of our leaving had never occurred to me during those last weeks, so crowded were they with people and doings; with the present. There had been no time to think of future goodbyes.

All at once I realized how much this place had grown on me. For, while Barnum hunted his elusive treasure at the end of the rainbow-rocks, I had found the end of my rainbow here in the village my pot-o'-gold in the hearts of these lovable people. Leave this fairyland with its wizards, its nats, its fables, its funny little customs, *my baby elephant?* It would be hard to think about—especially Bimbo. I couldn't bear to part with him.

Recovering from the first shock, I managed something about how nice it would be if we could stay on a bit longer. Barnum disagreed. "You know we have to move on, Pixie. My work's done in this region, and there's a lot of ground to cover between here and Monywa before the rains break. That's what worries me most. The rains. They're almost due."

So Barnum made plans to evacuate Gyat and march on to new worlds of discovery, leaving me pretty much in the doldrums; that

183

is, except for one ray of hope. There might be some way of taking Bimbo with us! Why not?

The more I thought about it, the more the idea appealed to me. But would it appeal to Barnum? There was the rub. Better bide my time and wait for the psychological moment when he's in the right mood, I counseled myself.

On our final night in Gyat I was still biding my time—what was left of it. We were seated on the *zihat* porch—the *tedje*, talkative as usual; the silent *pongyi* (monk); Barnum and I—attending the obsequies of several weeks of delightful living that all too soon would become a memory. Last things were being said, last looks exchanged. During a lull in the conversation, I breathed a "now or never," and the question nearest my heart popped right out: "What about Bimbo?"

My husband glanced at me inquiringly.

"I want to take him with us," I said.

For a moment he massaged vigorously behind his right ear, then laughed aloud. "Surely you're not serious."

My heart sank. "Yes I am. I want to take Bimbo home."

Barnum drummed his fingers on the chair arm, smiled charmingly for the benefit of the guests, and again regarded me long and hard. "No," he said rather tonelessly, "Bimbo stays here. I'm not having any pet pachyderm tagging after us the rest of our lives. Where you going to keep him? In a New York apartment? A hotel suite, perhaps? You know he's likely to grow into something too big for anything but a ranch." He lighted his pipe, meditatively. "There'd be the zoo, of course, but—and, moreover, it would cost several hundred dollars to ship him to the States."

"Only three hundred," I corrected. It was no use. Barnum simply set his chin the firmer.

"So, we just up and leave him," I chided, with scathing eloquence. "That's easy enough when you're dealing with your fossil variety of elephant. The living kind are different. They have a way of endearing themselves to you . . . but, then, you wouldn't know about that. You're too busy digging up the dead to have any time for the living. I'll bet you wouldn't think of leaving him behind if he were petrified and several million years old."

My worthy opponent shot me one of those I'll-get-you-later looks. "Had we kept all the pet monkeys, mongooses, mice and dogs you took a fancy to, we would have had to charter an ark long ago."

But I was in no mood to be brushed off. Here was one mother who did not intend giving up her baby without a fight. A few more jabs like the last one, a left hook to his pride, and I'd have him on the ropes.

"Why, I'm amazed that you—*you*—would even suggest going on without Bimbo," I rattled ahead. "Suppose you had been presented with a live baby dinosaur. Would you leave it? We have a cart full of fossil elephant bones—the lot of them old, cracked and busted—yet, you have seven fits if someone so much as breathes in their direction."

Barnum coughed, opened his mouth to say something, then closed it abruptly. I could see that he was resigning himself to fate.

To cinch the matter, I threw in an extra punch. "Here we have a perfect specimen of living elephant—skeleton fully articulated, not a bone missing, plus all the vital organs not only in a state of perfect preservation, but *functioning*. And what happens? You don't want it. You'd rather hunt your head off for a few broken fragments of some antique pachyderm that gave up its ghost way back before the Ice Age."

My husband's hand shot up.

Ah! surrender, I gloated to myself. But too soon.

Barnum was just dealing an ace he had up his sleeve. "Very well," he sighed, making a great show of relenting, "tell you what we'll do. We'll leave it to the man best qualified to render a decision—the *tedje*." He turned to the magistrate who was puffing away on a cheroot, obviously enjoying the performance.

Bimbo, awakened by the loud talking, had emerged from under the porch and stood peering up at us. Sleepy-eyed, his skin falling in loose folds about him, he looked for all the world like some grotesque dwarf lost in a pair of over-size pajamas. He seemed to sense something. His small trunk arched upward, curling nervously; his ears twitched. I could almost read the unspoken question in his eyes.

I glanced about at the three men. How strange that we should be sitting here in this far land, every fibre of our beings vibrating to the issue of what to do with one tiny elephant?

The *tedje* was slow to speak. When he did, his smile faded and the words came reluctantly. "Sahib right," he said, talking more to Bimbo than to us. "Child elephants not ver' strong. Not living long if going away. Better staying here."

The man hesitated, searching my face through the haze of cheroot smoke, then went on, "But could taking chance. Living in zoo maybe alright . . ."

What to do? It seemed silly to become so worked up about an animal. A baby elephant has a way of getting under your skin, though, and when you try to pull him out, part of your heart comes, too. I had Barnum practically saying yes. What more did I want? It was up to me.

"What will become of him if I leave him?" I asked, steadily.

The headman smiled reassuringly. His arms spread wide as if to encircle something huge. "Bimbo growing big and strong. Some-day working in logging camp. This ver' good—plenty food—plenty play. Elephant all same as man to Burmans."

A sudden choking hotness welled up inside me, blurring the figures of the men. Desperately I tried to erase Bimbo from my mind, to think of something else. Then—an overwhelming tired-ness swept over me and I wanted to be alone.

When my thoughts came clearly again I was lying on my cot, the pillow warm and wet against my cheek. The thing was still there in my throat, dry and bitter, and I could hear voices coming from far away.

I listened. The *tedje* was saying something about Americans always running after happiness—chasing it, striving to possess it, thinking happiness was property and owning things, and, because of wanting it so hard, not finding it.

I heard him talking about his people, and how happiness for them was in the enjoyment of things they had—their trees, their flowers, their hills and lotus ponds. A wife singing over her daily chores. The friendly voice of a neighbor. The sounds of home and baby. All these things his people enjoyed no less than the deep restful content inside a man's heart, and the sharing of what he had with others.

This, I thought, is where my little pal really belongs; here in this joyous land with these dear people who never seem to grow old; where man and beast awake each morning to a world bright and fresh and new; where rivers sing, trees have souls, and each unfurling flower forms the throne of some phantom fairy queen.

The *tedje* was right. Bimbo stays. And, somehow, the crushing heaviness was gone and I felt quite happy.

I have never regretted leaving him, though each year when spring comes 'round again and I can almost smell the *frangapani* and the *champa* my heart goes back to Burma and to Bimbo. I think I hear the tinkling of his small brass bell, and I fall to wondering how my baby has turned out.

He'd be a big boy now— "a-pilin' teak in some sludgy spudgy creek,"—the boss-elephant, too, I'll wager—a fate better by far than would have been his as a circus performer or a do-nothing dumbo in a zoo.

# 35
## ROYAL SPIRITS

Ahead was Mandalay. What visions I had conjured up about this romantic city of Kipling! There would be one week, maybe two, of luxurious living at the finest hotel, with nothing to do but relax and enjoy it. This would mean sleep till noon—brunch in bed—carriage to see the town—dinner with all the trimmings—tickets, front row center, to the best show—after-theatre party at the British Club. And, because of my happiness, there would be two weeks furlough for Mari and Dos.

Hallucinations of grandeur! That's what these dreams were—dreams hatched under a topi in the broiling sun. Delusions born of saddle days and camp-cot nights, and mirages in the jungle. Well, the dust of Mandalay was a good place to bury them. And I did—as soon as we entered the city.

Any romantic illusions left were in strict disregard of the facts. The town's one and only hotel was "full up" with the three R's—rats, roaches and ruin. We betook ourselves elsewhere, and fast.

Elsewhere turned out to be a neat Dak bungalow beyond the city-limits, in a grove of cool *neem* trees, with a wonderful well, private bath, and a corps of trained servants. Delightful vacation spot, I thought, not to mention its possibilities for honeymooning.

There's nothing like a few months in the jungle to sharpen one's appreciation of the comforts of civilization. Merely being settled somewhere felt good. Those last weeks on safari had been trying, what with racing the rains to Monywa and prospecting at the same time.

Our hunting hadn't been bad—considering. That *Anthraco-there*, the rare pig-like specimen Barnum had bagged, was worth the trip. True, he found only the skull, but it was a fine one, beautifully preserved and of a type hitherto unknown. The jaws and teeth of an old rhino-like *Titanothere* weren't to be sneezed at either, nor the batch of ancient turtle bones I had located in the walls of a gully, nor the many alligator vertebrae we credited between us.

Something petrified filled every inch of cart space and at one village we had to jettison part of our supplies in order to stuff in more petrified bones.

Our pockets bulged, too. As I remarked to Barnum at the time, "If you expect me to hold anymore, I'll have to open a pouch in my paunch. What am I—a wife or a kangaroo?" Even my saddle bags carried their quota of prehistoric sundries—a rib here, a backbone there, teeth, toes, sections of tail.

Small wonder the natives fled in terror when we rode into Monywa. We were a ghoulish-looking lot. They probably thought we'd rifled all the graves in Burma and were packing off the skeletons to grind up into some white man's magic medicine. But a few rupees quieted their fears sufficiently for some of them to help us crate and load our treasures at the railway station, whence they were dispatched to the American Museum.

Glad to be rid of the things at last, my husband sighed a final, "That's that!" I could only add that I felt somewhat like a fossil myself and wouldn't mind being nailed up in the last coffin, labeled *Lilianus brownus*, and shipped somewhere WEST of Suez.

At this the scientist in Barnum rose to the surface. "And crowd my specimens?" he laughed.

We presented a perfect picture of "love in a cottage" for two days, most of which time was spent at the typewriter, with Barnum in a dictatorial mood, spelling out long words to add to my growing vocabulary.

For more than two years now I had been exposed to his scientific double-talk, and some of it was actually beginning to make sense. For instance, during this particular session I learned that the rainbow-rocks we had followed through the jungle were forty or so million years old, and that we hadn't found any dinosaur remains in them because they were not quite ancient enough. Minimum age for dinosaur beds: sixty million years.

But, aside from the interesting bits of geologic fact one picks up, being stenog' to a bone-hunter is no picnic. After grinding out words like "Palaeozoic," "Mesozoic," and "Cenozoic," and phrases like "anterior extremity of the ischiatic process," "ascending ramus of the left mandible," and "phylogeny of the Ungulata," one is only too anxious to change the subject.

So, with the final "i" dotted and the report banished to the desk drawer, "How's for a spot of tea, darling?" I suggested, alluringly. "And after dinner, the movies. And tomorrow a tour through the Royal Palace—and—next day a ride up Mandalay Hill to see those wonderful pagodas the guide book tells about."

Silence.

"Well?" I urged, knowing from his sheepish expression what was coming.

"Fact is," he said, "I've just picked up a lead on some vertebrate stuff across the river. It'll bear looking into . . ."

That was enough for me. Why remind him that he had promised to rest and see something besides bones for a change?

"Take Mari with you," he finished lamely. I knew I'd have to see Mandalay sans husband.

Mari wasn't well next day, so I lolled alone in the old battered *gharri* carriage, feeling highly luxurious and longing for someone to share myself with. It was a beautiful morning, the flaming blossoms of the *gulmohar*, or peacock tree, scenting the air, and temple bells softly calling the world awake. On the outskirts of the city neat little houses lined neat little streets. Barefoot boys, wearing a combination of daddy's shirt, Mamma's skirt and baby's crocheted cap, scurried off to school. In the distance I could see Mandalay Hill, its myriad pagodas gleaming in the sun, and the rolling Shan Hills edging the horizon. Nearby were the throngs and confusion of mid-town; then—the high stone walls, the moat, the peaked roofs and towers of the Palace.

The Palace was not a particularly impressive affair. Huge, rambling, trimmed with all manner of grotesque architectural bric-a-brac, it resembled most other Oriental buildings, except that there was more of it. Nor was the place apt to inspire one with its cheerfulness; it had that "old" look—a sad, haunted air assumed by dwellings when nobody lives there anymore.

But, putting my imagination in working order, I felt a tingle of excitement as the carriage drew up before the main gate. After all, it wasn't every day that a girl had a chance to hobnob with royalty. Of course, they were only the *ghosts* of royalty, but that was an advantage since it's much easier to gain an audience with a dead king than a live one. And who ever heard of a spirit, however noble, slapping anyone into a dungeon simply for being short on court etiquette? Ghosts are nice people—especially those with blue blood in their past.

To slip through the cordon of guards at the Palace entrance would have been worth my life, a hundred years ago. Even had I been lucky, I couldn't have crashed the solid ranks of spittoon-bearers nervously awaiting the royal cud. I now brushed by all these ghosts without so much as a "Halt! Who goes there?"—my passage disputed by nothing except two old muzzle-loading cannons whose shooting days were over.

Within the building, vast gloom, emptiness, utter silence! When my eyes had grown accustomed to the dark, I found myself in the Main Audience Hall. Its far end, on a raised dais, held the seat of the mighty—the Lion Throne of gilt and glass not looking quite so "ritzy" as when the Burmese kings sat in glory while their subjects bit the dust.

Two carved doors opened onto the throne platform, on one the image of a peacock, on the other a hare—twin symbols of royal descent. This is not to say that the kings of Burma had rabbits or birds in their blood, these two creatures in reality being the Sun Goddess and the Moon God in disguise. Fortunately, they were of opposite sexes, as, otherwise, Burma might never have had a royal family. It is said that on the day "delivery" of the first ruler was expected, sun and moon came together in heavenly conjugation—

to put it unromantically, in eclipse,—and a king was born. So much for the divine origin of kings in this now-democratic land.

I was somewhat disappointed upon closer investigation to find no mystery to the double doors—not even secret passageways. Aside from a certain decorative value, their chief purpose was to permit a hasty get-away in case the royal person had to run for his life, which was likely almost any minute.

Being a king in Burma had such occupational hazards that it became traditional for a new monarch, on ascending the throne, to liquidate all members of the previous court, and do it thoroughly enough to insure himself a tranquil reign.

Kings practiced this quaint custom with variations right down to old Thibaw, last of the line. Standard procedure was to "give them the axe," as beheading was called in polite society. But Thibaw was a rare kind of king; he had ideas of his own. Individual decapitation not only was messy but a flagrant waste of time. A cleaner quicker method had to be devised.

Thibaw finally decided to toss the opposition party into a large hole and cover them with earth, the burial trench being dug just outside the Palace walls where the souls of the departed would provide a sort of ghostly watch for the royal precincts from then on.

The doomed did not depart quietly, however, the guards reporting some time later, "M' Lord, the ground moveth up and down." Upon which Thibaw, never at a loss for ideas, ordered the imperial elephants out for a walk—back and forth over the heaving earth until it was still.

When the king wasn't in his Lion mood, he had a choice of other thrones—the Deer Throne, the Conch Throne, the Elephant Throne. Who did what on the Goose Throne is anybody's guess, for while searching for the goose I fell through the floor. This gives a general idea of the condition of the royal tinderbox.

A long corridor with a railing of empty green bottles led from the Goose Throne. This was the Morning Levee Room where His Majesty conferred with cabinet members. From the condition of the railing, a good time was had by all.

Atone end of the council chamber, handy for eavesdropping, a door opened into the Chief Queen's apartment. Called the Glass Palace, this was a sumptuous nook lined with mirrors in which the First Lady could get a good view of herself from all angles while listening in on hubby's state secrets. Her own personal business she transacted on the Bee Throne in these same quarters. All other private business was handled on the Lily Throne, conveniently situated in the Ladies' Room.

In spite of court gossip the Privy Councilors, familiarly known as the *Atwin Wuns*, did not hold secret sessions in the Ladies' Room, nor gild the lily on the Lily Throne. Their concern was primarily with the Elephant Throne in the Treasury which, at Thibaw's abdication, contained a measly five thousand pounds.

However, as wealth was reckoned in women rather than rupees in those dear deluded days, His Majesty probably did most of his bookkeeping in the harem. At least, that was the case with Mindon Min, Burma's most benevolent ruler.

Now, Mindon Min was a gay old blade with a taste for the ladies. Luckily, he lived at a time when he could indulge that taste and call it his duty.

Large harems were politically expedient in his day. Just as, in our own era, a government's standing is measured by the size and tonnage of its navy, so in Mindon's world it was the size and avoirdupois of the female establishment. Naturally, keen rivalry existed between the various Oriental Chiefs of State as to which could accumulate the largest, not only for the national prestige that accompanied it, but also for the obvious distinction it gave the person of the king himself. There were no limitations—and that's where King Min came a cropper.

Early in life he had set his heart on a super *zenana*. *Four hundred and fifty wives* was his goal. That he might keep accurate tally of the figures, he had erected in the courtyard a small stone monument to each little dream girl he hoped someday to marry.

As the years passed the sweet young things increased. The proud king now spent most of his time hieing himself to the women's quarters to fondle his treasures and count them, like

Midas. But, though the spirit was willing, the flesh was weak, and as more years passed all he did was count.

The aging monarch began to worry. He was still far from his goal, with only eighty wives down and three hundred and seventy to go. Worry didn't help, and that was where the score stood when Min gave up the game. He just didn't live long enough to marry, or even meet, one-fifth of his longed-for quota. In the courtyard the small white slabs still stand where he placed them—memorials to the wives of a king who counted his chickens before they were hatched.

Not far distant in the Palace grounds is the glittering looking-glass tomb in which they finally buried the old boy. It has become a national shrine, for although Mindon Min never achieved success in the family way, he turned out to be an able statesman and wise ruler, beloved of his people.

Set among the gardens, grottoes, lakes and fountains of the Royal City were umpteen other buildings: pavilions for the junior wives and princesses, and for the amusement corps of dancing girls and mistresses; quarters for the "also rans"; the nurseries; the Royal Theatre of glass mosaic; the Tower of the Tooth, said to contain a relic of the Buddha; last but not least, the stable of the Lord White Elephant.

The sacred white elephant was an important part of His Majesty's household. Honored, pampered, he was the pride and joy not only of the court, but of the common folk. To the Burmese a white elephant was much higher in the scale of existence even than a human.

They are a very rare and delicate animal, hence every precaution was taken to keep the beast happy and healthy. As an infant, it had a special brigade of breasty females kept on tap for its milk supply. These women were reserve stock. When baby felt the need for nourishment, an emergency call was rushed to the milkmaids who stood by, ready to serve at all times. The first recruit would salute, execute a smart about-face and waddle over to the nursery which, no doubt, was fast being reduced to rubble by the impatient brat. After she had done her duty, the next performed, and so-on down the line until the first platoon was drained. The second platoon then stepped into the breach for a repeat engagement.

At bedtime the tiny tusker was first bathed in scented sandalwood water by a score and more of attendants, among them no less a personage than the Secretary of State, then cooed to sleep by a choir of soft-voiced singers. No wonder the Lord White Elephant died when the British shipped him down to the Rangoon Zoo after the capture of Mandalay!

This brings us to the taking of Upper Burma in 1885 when the British walked into Mandalay and took over the town—lock, stock and Palace—without firing a shot. Some have called it a military operation; if so, the name should have been, Operation "Picnic." General Prendergast, Colonel Slayden and their men merely started playing with some maneuvers in the direction of Mandalay one morning and by the time they reached the city it was theirs.

The General was a gentleman, however. After all, it wouldn't be cricket to snatch the throne from under the royal bottom without due formality. So from Ava, a few miles south of the city, he dispatched to the King an invitation to abdicate. It took the form of an ultimatum, of course, and touched off a squabble as soon as presented.

This was the only real fight of the campaign—a battle of the sexes between king and queen. He was all for submitting, she for holding on. The man, as usual, had his way.

What else could the poor king do? With the British Lion roaring at the gates, the Peacock and the Hare of Burma were in somewhat of a jam. Colonel Slayden, twirling his swagger stick, and smiling the imperial smile that has won the English so much, strolled into the Palace, stayed for dinner and spent the night—a night sixty-three years long.

## 36
### Bazaar in Mandalay

I had hoped secretly that the reported "find" across the river would prove a dud. Barnum needed rest badly, and I knew he never would get it so long as there was a bone anywhere around.

All my hopes were dashed when he flew in the door that evening with the glint in his eyes that I had come to know so well.

"We've got something!" he bubbled. "Just a few bones showing, but they look good, Pixie. Awfully good. Putting up a tent tomorrow. I'm giving the place a thorough going-over."

"When do we leave?" I asked.

He turned with deliberate indifference and sniffed the air. "What's for dinner tonight?"

"I repeat, when do we leave, Mr. Bones?"

Barnum had quietly maneuvered into the other room where I could hear him busily rattling papers.

"What in heaven's name are you so cagey about?" I called out.

His face, wearing its sad-little-boy look, peered from behind the door. "I'd much rather you kept things going at this end of the line," he admitted. "There's the report to type out, that batch of letters to get off. And someone's got to be on hand to call for mail in Mandalay. I'm expecting word from the Museum advising us on future operations. I'll only be gone a couple of days, Pixie."

"I know. Only your days have a way of stringing into weeks—or even months."

"Anyway, I'll be home for the week-ends," he added, by way of appeasement.

The report and stack of letters looked entirely too formidable the following morning. The day was much better for shopping. I grabbed Dos and we were off.

I directed him to make a bee-line for the silk stalls. The Mandalay mills were famous for their lovely iridescent silks, and I had my heart set on buying some.

But who can keep her mind on any one thing in the great Zegyo Bazaar, with its maze of shops selling just about everything under the sun. Spice shops. Toy shops. Lacquer-ware shops. Perfume shops smelling of sandalwood, musk and jasmine. Shops selling jade and semiprecious stones cut into rings and chains. Booths where silversmiths and goldsmiths sit tailor-fashion all day long, hammering, carving, shaping precious metals into curios, bowls, trays, and jewelry for the ladies. Shops of fine Burmese amber. Gem shops where, by appointment only, they will show their gorgeous sapphires, amethysts, and rubies—the pride of Burma.

I passed through winding lanes flanked with footgear of all kinds—sandals of plush, leather, wood; some trimmed with beads; some turned up at the toe, others down at the heel. Over all came the shrill cries of women calling out their wares, Hester-Street fashion.

The nose knows when you reach the food mart. That Limburger odor can be traced to mounds of *Ngapi* fish paste rotting in the sun. It's rivaled only by the fumes from the large screened meat market where women sit atop counters of lamb, goat and pig, shooing off the flies with switches of false hair. When business is slow and they feel a nap coming on, they grab a side of this or a leg of that and siesta on the spot. Naked children play about. Every woman has at least two, one swinging on the meat, another cradled in a basket hung overhead out of harm's way. The crawlers are getting their first taste of life in the raw on the floor.

Women run the bazaar not only with brains but with beauty—fresh and lovely as the flowers in their hair. In the silk shops the vendors are all young women. It pays to advertise, as witness my pretty salesgirl sporting a goodly array of finery on herself. Her sales technique is deceptively casual, and she has a dainty way of doing you out of your right eye in the bargain. Reclining full-length on a matting rug inside the booth, coiffed head resting on a wooden pillow, she nonchalantly puffs a perfumed cheroot while displaying her wares and her charms—which informality puts the customer at ease and in the mood to buy.

But how was I to choose from the stacks upon stacks of dazzling silks that shimmered with all the colors of the rainbow? Had it not been for Dos, I'd still be there.

The narrow lanes of the bazaar swarm with strange people from every nook and corner of Burma—Shans from the jungle; Karens from the nearby hills; Chinese from the border; erstwhile Padaung headhunters; friendly Kachins. Each displays his own distinctive dress.

Padaung women concentrate on the necks, these stretched out to accommodate high collars of metal rings. They look like so many animated carafes in which the wine has turned to vinegar. But they could hardly be expected to look sweet with sixty pounds of solid brass weighting down arms, legs and other spare parts, in addition to a metal "handle" on the back of the collar for friend husband's use when occasion requires.

With the Kachin girls the accent is below the belt. They wear bright woven skirts with dozens of bamboo hoops around their middles denoting their wealth, and black velvet jackets trimmed with silver dollar buttons. Some have gay woven leggings. None are bothered with bootery. Large silver cuffs and necklets set off the stunning costume.

I was so taken with the Kachin outfit that I determined to have one. I singled out the best-dressed girl of the group and, with Dos interpreting, offered to buy the clothes off her back.

My proposal evoked a volley of laughs, accompanied by suggestive glances between Dos and the other salesgirls. Mari would have snatched him bald-headed had she been there.

"What does the lady say, Dos?" I asked. "Will she or won't she part with her clothes—naturally, at a more convenient time?"

"The lady say *No!* Not wanting to go naked. She having only one skirt, and taking two years to make new one."

We had another round of laughs and I considered the matter closed. How was I to know that she would turn up at the bungalow some days later, prepared to part with her prized costume—for a price?

# 37
## Wanted—a Miracle

Barnum didn't know it yet, but our days in Mandalay were numbered. A cable from the Museum containing our marching orders had settled that. I was busting to break the big news to him when he returned from the dig that week-end.

However he was busting with news of his own—and his wouldn't wait. "I've uncovered two of the sweetest elephant jaws you ever saw," he announced, jubilantly. "A couple of first-class palates, too. This river bank has the finest outcrops thus far in Burma. When I finish at this spot, the entire series of beds both up and down stream are in for some honest-to-goodness prospecting. Why, I'll wager—"

Something in my manner brought him up short. He gave me a searching look, then asked abruptly, "Anything happen while I was away? No secrets now—out with it."

Slowly, deliberately, I lit a cigaret, rested my head on the back of the chair and inhaled long and luxuriously, while I watched my husband through the blue haze. This once I happened to be on the other end of the news. It was a rare occasion and I wanted to linger over each fleeting instant of it. Too many times had I been kept in the dark about our next move. Now it was my turn.

At last I relented. "Would it interest you to know," I queried archly, savoring each word, "that you will be doing no further prospecting in this area?"

"Wh—what?"

I continued. "Other plans have been made for you in other places. As a matter of fact, we're on our way to—China. *Just as soon as we can pack.*"

His baffled expression changed to one of complete bewilderment.

I held out the cable.

He grasped it and devoured the contents, reading aloud— "Plans made for you to extend expedition into Yunnan, China. Reports of unusual geologic finds there. Leave soon as possible. Detailed instructions mailed to Bhamo."

He looked up, eyes sparkling with excitement. "Just as I expected," he said, trying to cover his elation. "Had a hunch we'd be getting up that way. Wouldn't be surprised if we hooked up with Roy Andrews in the Gobi."

"Gobi, here we come!" I shouted, our week-end breaking up in an orgy of fun and laughter that had the house servants goggle-eyed. Barnum rushed back to finish work across the river while I plunged into packing and preparations for the long journey to Yunnan Fu.

That's as far as we got. Next thing I knew, Mari took to her bed with fever. Four hours later she was in the hospital. Two days later she was out again—no one knew where.

According to the hospital, she had mysteriously disappeared during the night, escaping in one of the establishment's best sheets. The nurse turned in the alarm when she discovered that what had been the quietly sleeping Mari a few short hours before, was an ingeniously-wrought facsimile made of pillows and rolled-up bed-clothes.

Thinking Dos would be the first to learn of Mari's whereabouts, I hurried to question him, only to find that he, too, had vanished into thin air. Strange! This was not like Dos. My anxiety grew as afternoon lengthened into evening. Night fell. Still no word.

My vigil came to an end shortly after dark. All at once Dos stood in the doorway—haggard, silent, his mouth working to form words that wouldn't come.

I moved close. "What is it, Dos?" I asked, watching his lips.

In a barely audible whisper, he answered, "Memsahib, come quick. Mari here."

We hastened to the servants' quarters. There I found my maid stretched across her cot, flushed with fever and moaning softly.

Hearing us enter, she rolled her eyes in my direction and asked, "Madame angry that Mari running away?"

"No, darling. But you shouldn't have. Why didn't you stay in the hospital until you were well?"

The sick woman sat up, bright-faced. "Me shopping for funeral clothes," she said cheerily, motioning Dos for something.

"Funeral? What are you talking about?"

Mari ignored my question, took the bundle Dos handed her and from it drew a bright new red *sari*. She held the garment against her, caressing the rippling silk. She smiled. "Madame like?"

"Mari," I said, "you must get out of this frame of mind. I'm going to arrange to have you return to the hospital tomorrow."

The smile faded from the fevered lips, and the pain came back into her voice. "Hospital no good for me, Memsahib. Me not getting well. Me going to die."

"Why do you say that? It's nonsense!"

"It is the rule, Memsahib. It is the rule."

"It is not the rule for young people to die. We can make you well again," I soothed, stroking her head.

She rolled from side to side on the pillow. "No, no. Me going to die. Dos taking me home for funeral . . . home . . . home . . ." She rambled off into incoherence and fell asleep.

From the shadows came the sound of Dos sobbing. I went to him. "This talk of dying isn't good," I said. "Mari is not going to die. She can be well in no time—only she must go back to the hospital where she will receive proper care."

Dos stiffened. "No!" he snapped, almost savagely. "Mari speaking truth. She going to die. No one can help. Medicine no good. She happy. Spending whole day in market, buying best funeral *sari* in city. She my woman. Me taking home. Giving fine funeral. Mari . . . she . . ."

He broke into another flood of tears. There was no reasoning with him. He, too, being a Hindu, fatalistically believed that his wife's time had come. Try as I might, I couldn't persuade him to think otherwise.

Compromise was the only way out. "Alright," I declared. "If you think that's the thing to do, I'll return with you and Mari to Yenangyaung, but not until she's had a few days' rest, first. She's in no condition to travel at present. But there'll be no hospital, Dos, I promise you that. I'll take care of her myself right here."

Dos nodded agreement. I prayed for a miracle to happen.

Whether those prayers would ever be answered suddenly ceased to matter. The miracle so desperately needed now was not for my servant, but for my husband.

# 38
## 106.3

When Barnum returned from the dig he complained of extreme fatigue. His face showed a strange pallor which alternated with hot flashes.

Somewhat alarmed, I put him to bed immediately and went into Mandalay for a doctor. At four in the afternoon I found one in the middle of a squash game. He accepted my announcement of Barnum's condition quite casually. "Touch of malaria, probably," was his off-hand verdict. "All get it here, y'know. I'll run over in the morning to see him. Frightfully busy just now—toodle-oo."

I insisted that he quit his game and come at once. After prolonged and maddening indecision he condescendingly agreed.

He was standing behind my husband's bed as he drew the thermometer from Barnum's burning lips. Noting its reading, his suave, smooth-shaven face went ashen.

The thermometer read 106.2 degrees!

"He must have been carrying this around in him for days:" said the doctor, thoroughly alarmed. "It's too late to—"

"Too late!" I gasped.

"I mean," he added, "he's too far gone to move to the hospital now. We'll have to do what we can for him here."

"Is it really malaria?" I asked, trying to convince myself that it couldn't be.

The answer thudded against my ears: "Mrs. Brown," he looked me soberly in the eye, "you may as well have the truth so you'll know what you're up against. This is *black* malaria—the worst."

He took a bottle from his medicine kit. "Liquid quinine—three times a day—sixty grains a dose."

I gasped. A quarter of that dose would have sounded enormous.

The doctor's voice droned on. "It's kill or cure, you know. Follow instructions . . . may pull him through . . . do what I can . . . no nurses available . . . you'll have to go it alone." He gave a sick, hollow laugh. "Feel up to it, Mrs. Brown?"

Each morning after the long, hard fight of the night, I'd drag myself into a hired *gharri* and drive into the city to its one and only soda plant. Here I stocked up on enough ice to last until next morning, keeping it in sawdust and burlap until the fever began to mount at six in the evening. Then I packed him in the ice, where he remained until two in the morning when the fever reached its height and his temperature commenced to drop.

I lived mostly on hot strong tea during those trying days when only the white-walled room and the long recumbent figure of my husband were entirely clear in my mind. Everything else seemed to move in another world from which came small sounds—the beating of a palm frond against the window—the drip, drip of rain—the mournful chanting of house servants outside.

The ghastly routine went something like this: Six o'clock, fever rising . . . 103 degrees . . . 104 . . . icepacks . . . *pray*. Then the long watch through the night . . . delirium . . . fever . . . 105 . . . 106.2 . . . 106.3 . . . life hanging in the balance of a half-degree. *Pray*. Then drenching sweats when the fever declined, and then—the long sleep so close to death that I often thought my darling had slipped away, only faint moisture on a mirror held to his lips showing that he still lived.

And sometime during those endless ages, Mari and Dos left for their home—where Mari was to die.

The house servants couldn't be depended on for anything, so I was left to battle it out alone. Outside my door day after day they continued their mournful chanting. The Sahib was doomed, they said, in effect, with typical Oriental resignation. If he was to die, he was to die; no medical treatment could alter the course of fate.

Not one of them moved a hand to help save him; rather, they were already planning the disposal of his body, there being no dirth of helpful hints in that direction.

They agreed unanimously that cremation was the thing. "However," I was cautioned, "when Sahib pass into the Great Peace, do not put him in closed coffin like they did General Z—Sahib. It took him six days to slowly roast before he turned to ashes. You burn Sahib on open funeral pyre of special sandalwood, as is proper."

"But the Sahib is not *going* to die," I kept telling them, gritting my teeth.

They only smiled and nodded their heads, repeating, "Sahib die . . . Sahib die . . . Sahib die"; the words slowly grouping themselves into a dirge.

In his delirium Barnum seemed to relive his past life. I strained to catch the words, hoping to learn something of his suffering. Much of his talk was a ragged babble, but now and then it grew even and distinct, and I could piece together the meaning.

It was mostly of his youth that he spoke. Sometimes he was a small boy again, wandering over the coal measures on his father's Kansas farm, collecting his first precious specimens. These fossil sea-shells and plant impressions that he proudly exhibited in the parlor when company came, soon filled all the extra rooms in the house and had to be relegated to the barn.

Again he was a student at the University of Kansas, studying geology under his much-loved Professor Williston. Then he and his classmate, Elmer Riggs, were in the field, digging fossils in Wyoming with the illustrious Henry Fairfield Osborn.

At Como Bluff Barnum, it seemed, had excavated the first dinosaur skeleton collected by the American Museum.

At other times, following the sequence of his early years, came details of his shipwreck off the coast of Patagonia, clinging for life to a broken hatch cover on which he floated ashore.

Finally, the words would turn into a meaningless whisper that ended in silence, or suddenly jumble together and be lost in his ravings.

Waves of terror engulfed me at such times. It swept over me full force that here I was facing my ordeal in utter loneliness—no one to turn to for advice or encouragement except my coldly-professional doctor. The natives were worse than useless; far worse because they had Barnum already dead. Whenever I dozed off for a minute or two my sleep was cluttered up with the most horrible nightmares.

Five days, five nights, the battle for life went on, with Death standing at the foot of the bed—poised and waiting. Five days and nights, measured in pulse beats and the rise and fall of a thin red line in a fragile little thermometer!

On the fifth night the fever broke. Said the doctor, "Now we can move him to the hospital."

Barnum opened his eyes. His lips moved. I bent low to catch the feeble whisper,

"Just tired, Pixie. Just tired—that's—all. Day's rest—fix me up," he ended, laboriously, little dreaming how close he'd been to the Great Divide.

When he came out of his delirium, we found that he couldn't hear a thing. The enormous doses of quinine had left him totally deaf—fortunately, only temporarily.

Crisis passed, no sooner did we move him to the hospital than the doctor found me out on my feet—roaming aimlessly about with a 104 degree temperature, and jabbering a lot of rot. The strain and worry and constant fight had finally taken their toll. It wasn't malaria in my case, though, but a rather common form of jungle fever usually brought on by exhaustion. For a time we lay side by side on adjoining cots, neither one knowing whether the other lived. My fever passed quickly and I was up and around, able to tend Barnum's recovery.

His "day's rest" lengthened into weeks at the hospital. Months followed at a British resort town in the cool Maymyo Hills, about 40 miles east of Mandalay on the road to Lashio.

Our inn, "Lizette Lodge," was the home of an English lady named Gertrude Routleff, whose East Indian cooking, adapted to western taste, earned her a wide reputation among gourmets. She closed the inn to all other guests while we were there, so that Barnum might have absolute quiet.

He had wasted to a mere shadow, and weighed less than a hundred pounds. For a long time he was forced to remain in bed. But we wheeled his bed out on the vine-shaded verandah in the daytime, where I read to him. He was too weak to do much more than turn over. His diet at this time was restricted to beef tea, and oatmeal soup with an egg yolk occasionally mixed in for variation. No solids.

After a few weeks of Mrs. Routleff's expert care he had regained strength enough to sit up in his pajamas; a few weeks more and he could stand by the bedside, bracing himself on my shoulder. Then— the first feeble steps, Mrs. Routleff supporting him on one side, I on the other. Finally, all he needed was a cane.

From then on we measured his recovery by the distances he could walk from the porch without fatigue, his strolls bounded by the edge of the garden. One day he made the rose bush; next, the rhododendrons. After that, he was walking clear around the house slowly, to be sure. But his step became firmer, his eye brighter, and he was actually laughing again.

Before we knew it, the man was dressing in his new Palm Beach suit and wanting to go places. Maymyo had quite a variety of attractions—polo ground, golf course, drives through the cool, wooded hills. At the British Club were squash and badminton courts, and a large swimming pool in which he spent many a recuperative hour.

One epoch-making day, when Barnum had pretty well regained his strength, we bade Gertrude Routleff farewell—a mutually tearful affair—and boarded the train for Rangoon.

Farewell to Burma—and to youth!

PART THREE

# AMERICA

"After the crashing seas, the harbor bar;
After the heated sands, the cooling well;
After the stones, the fields of asphodel;
After the storm-torn night, the morning star."
—Edwin Markham
(Reprinted by permission)

39

BROWN'S BAILIWICK

There is something overwhelming about coming into New York Harbor from the great emptiness of the Atlantic. It bursts upon you in a sudden rush of noise and feeling—a vast walled city rising from the sea. You don't know whether to laugh or cry. Barnum and I did both that night as the *Mauretania* moved slowly up the Hudson River to her mid-Manhattan berth.

It was New Year's Eve—the midnight hour, to be exact. The air was clear and crisp, and a winter moon followed us from behind the turreted shore. Tides of sound spilled from the land, blending with the immediate roar of the river traffic. We heard the chime of bells; the distant hubbub of auto horns; the hoarse croak of freighters; the wail of sirens spiraling up from fire-boats and the fast cutters of harbor police. Old New York was out to welcome us home, it seemed, and, as if in answer, came the *Mauretania's* deep booming— "Ho-o-ome! Ho-o-ome!"

Home! What a wonderful word, I thought, my eyes tracing the old familiar skyline. New York was home for both of us. We'd both been away for four long years! That was me out there—my city— my birthplace. I was part of it. Never would I leave it again.

Barnum took me in his arms, hugging till it hurt, and we laughed and babbled and shouted Happy-New-Years to everyone. Then we joined the crowd below where, amid a riot of colored streamers, balloons and confetti, we toasted the New Year in to the singing of "Auld Lang Syne," the rasp of tin trumpets and squeal of paper horns. What a homecoming celebration! Nor was there

211

an end to it when we went ashore, nor the next day, nor the day after. That wonderful crazy night continued on and on.

Weeks of indulgence followed, sating ourselves with all the things we had longed for while away. New York was a thing to luxuriate in, to feel, to smell, to taste. It was coffee the American way, ham 'n eggs, deep wedges of apple pie. It was loved ones. It was friends old and new. It was walking up Fifth Avenue, browsing in the shops, mingling with the throngs and knowing that you *belonged*. The theatre; movies. Cocktails at the Colony Club; steaks at Cavanaughs; a spiggoty Italian place in Greenwich Village. It was the feel of plush carpets underfoot, soft lights, music, dancing at the Plaza, the Ambassador. It was paradise! Soon our years in the Orient assumed the hazy outlines of remembered dreams.

Almost without knowing it (so easy was it in those days), we found ourselves with an apartment—a luxurious, sprawly suite west of Broadway and not too far from the Museum. Known to friends as "Brown's Bailiwick," the place looked darling with our many treasures strewn about—bright little reminders of faraway places to keep our memories warm.

There was the Kashmir lamp on the foyer table; the small stone Buddha from a Burmese temple sharing the top of the bookcase with a bronze likeness of the Hindu God Vishnu; rugs from Baluchistan; silver cheroot bowls brimming with roses; Mandalay lacquerware for cigarets; Indian jade for ash trays; turquoise mosaic work from the land of Lalla Rookh; a frieze of choice photographs high around the study wall.

We gave cocktail parties highlighted by a new drink named "The Brown Special," and served with swizzle sticks marked "Stolen from the Brown House." Perhaps a broiled chicken dinner would follow, with pilaf of wild rice or an Oriental curry creation that brought everyone flocking around the stove, the finished product ending up as some kind of international potpourri composed of nearly everything in the kitchen.

During this period of bliss my husband's domestic side showed itself. To my astonishment, not only could the man cook and bake, but put up jams, and bread-and-butter pickles, and mix the most divine salad dressing you ever tasted. Housekeeping? He loved it. Saturdays I had to fight him for the vacuum.

Among his intimates, a certain amount of gossip apparently had preceded me. News of his marriage in India had started a rumor to the effect that he had really hijacked his wife from the harem of a maharajah. Perhaps I should have played the fiction up instead of down, for undoubtedly a "haremite" would have been much more intriguing than a mere convent girl with a flare for writing. There were times when some of my husband's co-workers professed to believe me the ex-Maharane of Patiala, or the Sultan's daughter, or what have you.

Ever since that memorable day when Barnum took his nose from the grindstone on his father's midwestern farm to bury it in bones at Kansas University, the Museum had been the prime fact of his life. He kept regular office hours there now—nine to five— and I liked nothing better than watch one of our monsters "get the works" in the laboratory. Any old fossil, feeling its hundred million or more years, would find a few months in the American Museum's clinic just the thing to put him back on his feet.

The lab looked like the workshop of some mad genius in the throes of building himself a nice, grisly Frankenstein. Pots, pans, bottles, jars, rubber cups, cluttered the long work benches, mixed in with spatulas, knives, chisels, awls and fine dental tools. The odor of chemicals filled the air—the ripe-banana smell of ambroid; of boiling dextrin which, mixed with plaster, served as a mend-all.

Sundry dinosaur cadavers huddled in the corners—the packaged dead still swaddled in their white plaster wrappings, each numbered and tagged with an elaborate case history. On shelves along the walls, spread-eagled on tables, or slung from the high ceiling, were specimens from all parts of the globe: a small hooded dino' from Mongolia; a slab of Bavarian limestone containing the fragile bones of a flying reptile—a *Pterodactyl*; remains of an extinct Patagonian mammal; an amphibian hailing from the Red Beds of Texas.

I soon felt enough at ease in the company of Barnum's menagerie to call them by their first names. *Stegosaurus* was "Stegy"; *Brontosaurus*— "Bronto; and *Paleoscincus*, just plain "Stinky."

Here I saw fossilized evidence of some of the things Barnum had only hinted at in his disjointed fevered mutterings in Burma—especially his work with dinosaurs. His early infatuation for these gigantic creatures had ripened into a love so strong that he gave some of his best years to unearthing dinosaur skeletons in odd corners of western United States and Canada. As a member of the American Museum staff he had accumulated enough specimens to stock a large exhibition hail with mounted skeletons and several store rooms with spare parts—an accomplishment which led to his being made Curator of Fossil Reptiles (chief keeper of dino' bones).

It was natural, then, that on my maiden trip to the Museum he should escort me directly to Dinosaur Hall—trophy room of prehistoric big-game.

Entering this hall is like stepping into a vast catacomb. Propped about in living pose stand the naked dead—monstrous shapes of bone vestured in horror and cast in the image of a forgotten world. A chill remoteness surrounds them, as if they had belonged to another planet. Yet, when Barnum explained them, speaking as one would of old friends, they seemed to change and warm into life.

"My favorite child," he said, introducing me to the soaring might of *Tyrannosaurus rex*. "Found him out in the Hell Creek Badlands of Montana. Took me two summers to dig the fellow up and transport his remains, by wagon, one hundred and thirty miles to the nearest railroad. Most formidable fighting machine ever devised by Nature. Stands almost twenty feet tall."

Hard by stood the sixty-six feet of elongated reptile known as *Brontosaurus*, the "Thunder Lizard." "Tipped the scales at forty ton when alive," Barnum stated. "Stuck to a vegetarian diet, and was harmless as a kitten."

"Providing he didn't sit on you," I inserted.

He continued. "Just as *Tyrannosaurus* over there was the ranking flesh-eater of his day, so *Bronto'* represented the highest type of herbivore. Came from Como Bluff, Wyoming where dino' bones were so plentiful fifty years ago that an old hermit built a cabin out of them. After the *Bronto'* had been excavated, we spent six years preparing and mounting the skeleton."

Strolling further down the hall, neither of us spoke. An ageless silence seemed to enshroud that incredible assemblage once flesh and blood, now stony reminders of Nature's magnificent experiment with bulk and brawn. And suddenly it dawned on me as never before why my husband was so obsessed with his work. It was a great work. *He* had done this. The amassing of these prehistoric wonders had been chiefly his doing, and it was not a small thing.

Suddenly my eyes caught sight of a familiar name—*Colossochelys atlas*. Behind it, completely mounted, stood the giant tortoise we had collected in the Siwalik Hills.

"It's our turtle!" I shrieked.

Barnum grinned. "I was wondering when you'd notice him. He was placed on exhibition yesterday."

Nostalgic odors of fish glue swept me back to the village of Siswan and my wild ride for shellac. This was *our* specimen—the child of Brown & Wife. I felt warm inside and glad all over, for now I, too, had a share in the Museum. From that moment it became a personal and intimate thing.

I caught something of Barnum's enthusiasm for the past as he hustled off each morning to bury himself in prehistoric bones. But when he returned after a hard day's work to loll around all evening in smoking jacket, or linger through a quiet game of bridge, the Browns, globetrotters no longer, were living the lives of normal human beings—and loving it.

"No tents to pitch!" Barnum would gloat, smiling reminiscently. "No camels to unload! *No monkeys to get in your hair!*"

I'd be thinking of something quite different. In spite of numerous valiant tries and at least one "almost"—that dreamy houseboat on the Kashmir Lakes—we never yet had had an honest-to-goodness honeymoon. A cable from the Museum, or the discovery of some likely old bones, had always interfered. But none of those things could happen here. What was to prevent our taking a belated honeymoon now—one with all the usual frills?

My husband agreed. We considered conventional trips by land, sea and air. There were no real discussions. Anything I suggested was all right with Barnum. There was some thought of a honey-moon immersed right there in little old New York.

Our plans got nowhere until the day he came home early with that peculiar glint in his eye that I hadn't noticed since we quit the Far East.

For a time he wandered from room to room uttering inane remarks and evidently trying to suppress some deep emotion. Finally, I planted myself in his path.

"What is it?" I demanded.

Suddenly he swept me into his arms and blurted out, "Pixie, how'd you like to go on a dinosaur dig in Wyoming?" He hesitated. "Of course, it would mean postponing our honeymoon—but—"

I brushed this last aside as unworthy of the prehistoric dead and asked, elatedly,

"When do we start?"

# Coachwhip Publications

## CoachwhipBooks.com

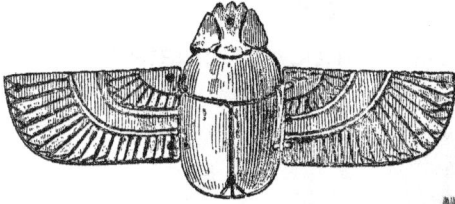

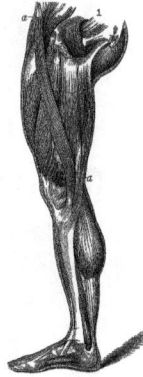

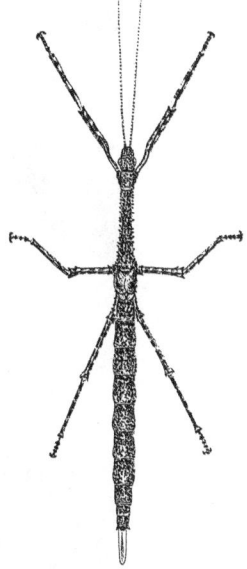

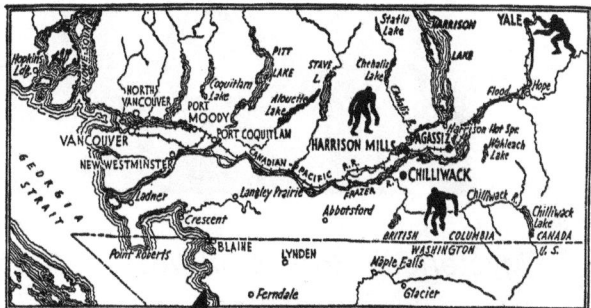

# COACHWHIP PUBLICATIONS

## COACHWHIPBOOKS.COM

Where did the Appaloosas and Pintos
of the northern plains tribes originate?

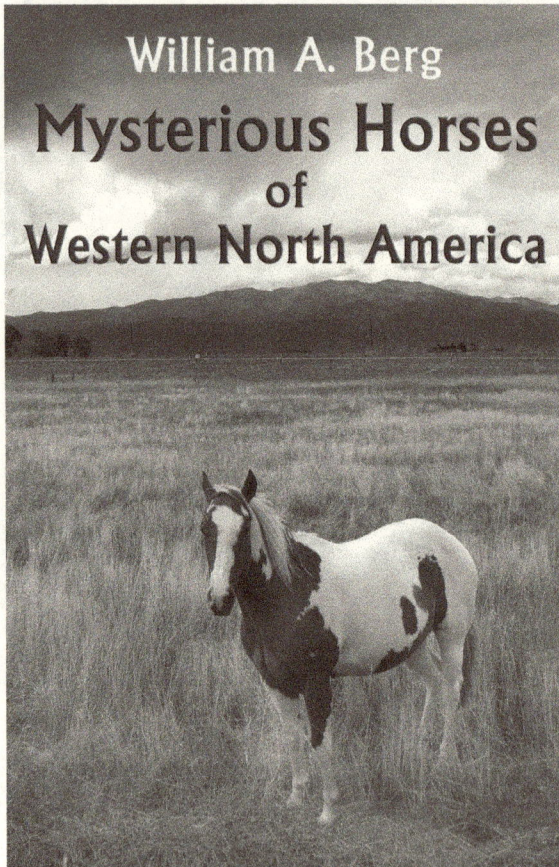

William A. Berg

# Mysterious Horses
of
Western North America

*Mysterious Horses of Western North America*
ISBN 1-61646-027-X

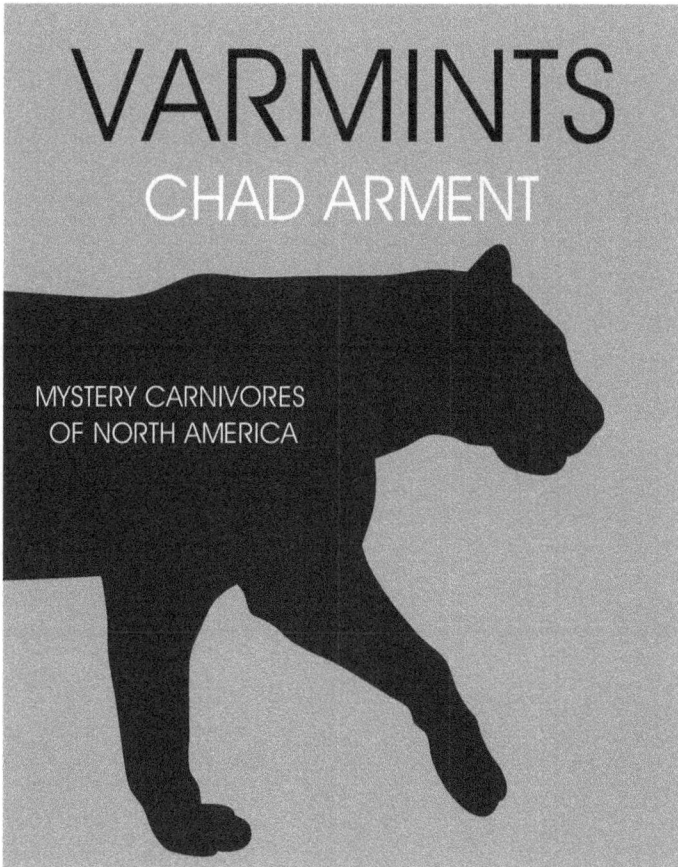

*Varmints*
ISBN 1-61646-019-9

www.ingramcontent.com/pod-product-compliance
Lightning Source LLC
LaVergne TN
LVHW011224080426
835509LV00005B/312